Salomon Jadassohn, Gustav Tyson Wolff

Treatise on Single, Double, Triple and Auadruple

Counterpoint

Salomon Jadassohn, Gustav Tyson Wolff

Treatise on Single, Double, Triple and Auadruple Counterpoint

ISBN/EAN: 9783337148911

Printed in Europe, USA, Canada, Australia, Japan

Cover: Foto ©berggeist007 / pixelio.de

More available books at **www.hansebooks.com**

TREATISE

ON

SINGLE, DOUBLE, TRIPLE AND QUADRUPLE

COUNTERPOINT

BY

S. JADASSOHN,

PROFESSOR AT THE ROYAL CONSERVATORIUM OF MUSIC, LEIPZIG.

TRANSLATED INTO ENGLISH

BY

GUSTAV (TYSON-) WOLFF,

MUS. DOC. CANTUAR.

LEIPZIG, BREITKOPF AND HÄRTEL.

NEW-YORK, G. SCHIRMER.

ENT^D STA, HALL.

1887.

PREFACE.

The subsequent Treatise contains instructions for the study in single, double, triple and quadruple Counterpoint. All the rules, principles and remarks set forth in this volume, are founded on the contrapuntal works of BACH, HANDEL, and other classical masters, who have written in our system of the major and minor keys.

These studies in Counterpoint are intended to prepare the student for the composition of Canon and Fugue; but those also, who do not intend to become musicians by profession, will be enabled to penetrate more deeply into the works of the classical masters and to cope with their sublime creations. Let no one imagine, however, that the knowledge of the rules alone would suffice; these would be attained quickly and with little trouble. Only serious, conscientious study can further the pupil, here, as well as in all other disciplines of art. Only then, when the student has mastered all the problems, contained in this book in a thorough manner, will he be enabled to proceed to the study of Canon and Fugue.

Leipzig.

S. Jadassohn.

CONTENTS.

PART THIRD.

PART FIRST.

Single Counterpoint.

—

CHAPTER I.

Equal Counterpoint.

§ 1. The term Counterpoint implies the independent progression of one, or more, melodious parts or voices, with, or against one another, taking into consideration a natural and correct connection of chords.

The characteristic feature of counterpoint is therefore melodious; each of the parts or melodies thus united, must be worked out independently; each must be an equally entitled part of the whole. This enables us to change at will, in double, triple, and quadruple counterpoint, two, three, or four parts or voices in their position, to, or against one another. Thus each part may become in its turn either soprano, alto, tenor, or bass.

We have already recommended to the student in the exercises in our book on Harmony, (where we dealt with the structure and connection of chords,) a greater amount of care and consideration, in the progression of parts from a melodious point of view. In the last exercises in the "Manual of Harmony" particular attention was called to the formation of bass and soprano. Referring to this, we can at once begin with the exercises in single counterpoint. We distinguish *Equal Counterpoint*, in which only notes of equal duration are placed to a cantus firmus, — and *Unequal Counterpoint*, in which notes of unequal duration are placed in one, or several parts against the cantus firmus. In equal counterpoint the progression of parts will be independent only with respect to melody; but in unequal counterpoint, the progression will be independent in melodious, respect as well as rhythmical.

The only difference then, between the exercises in equal counterpoint, and our last studies in the "Manual of Harmony" is this, that the choice of harmony employed shall be now free. By this means the opportunity is given, of bestowing especial attention to a more melodious progression of each individual part.

We commence our exercises as before in the four-part phrase, and place the cantus firmus in the bass, to which the student will have to find the three upper parts. He should treat these in different ways, with respect to position and choice of chords. It is intended that the student should employ for the first few workings of the exercises, only diatonic chords, and to choose for the beginning as simple harmonies as possible, and only, after he has done so, to allow himself the application of more distant or seldomer used harmonies. Then, after the cantus firmus has been worked out several times with diatonic harmonies, he will be allowed to employ modulations also. These however must not lead too far, nor be introduced in an unnatural, or forced manner. The treatment of the subjoined bass may serve as further explanation.

Cantus firmus.

Note. The student is recommended to work his exercises always in four different clefs. Here the following examples are noted on two systems, merely however, to save space.

At the above bass, only the six notes of the 2nd 3d 4th 5th 6th 7th bass allow a change of harmony; the chord of the first and last bar, must necessarily be the triad of the tonic; the chord of the last bar but one must prepare the close as Dominant. Nevertheless this cantus firmus allows a great number of different ways of treatment. The two first of the above examples contain only common chords;

in the third and eighth we find the chord of the Dominant Seventh; in the 4th 5th 6th 7th 14th 15th and 16th examples, diatonic chords of the seventh are used; in the 9th and 10th examples, we find the secondary chords of the seventh, of the key of *C*-major on the 2nd and 7th degree with the altered fundamental note and altered third; only the 12th and 13th examples give transitorily some modulatory progressions to the dominant of *a*-minor; the chromatic alterations being effected, of course, in the same part (alto) to avoid false relations. The cantus firmus would allow still different ways of treatment; those given here however, are sufficient for giving the student directions, how to work out his exercises.

It may be remarked here, that it is not necessary to give only triads to the first treatments of the following basses, as shown in Ex. 1 and 2.

Exercises.

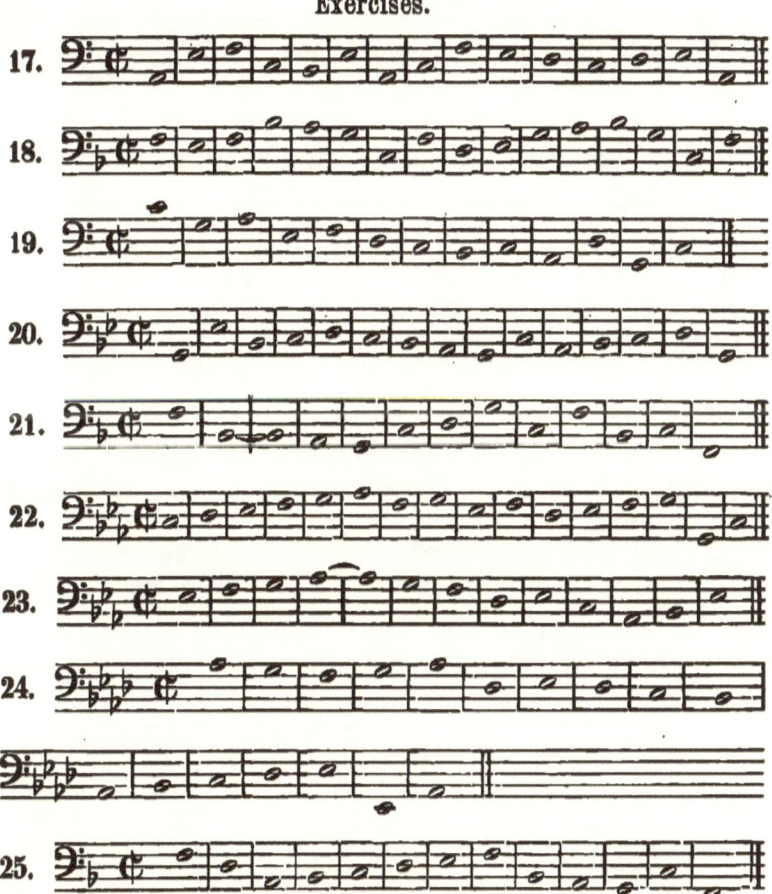

For the guidance of the student the commencement of the exercise No. 17 may be noticed.

The Cantus firmus in Soprano.

§ 2. We give now a cantus firmus in the Soprano. Here the progression of the bass will require the greatest care. (See Manual of Harmony § 61.)

No new rules are required; as practical guidance, we give here a few examples of the commencement of the cantus firmus.

Exercises.

The exercises No. 30. 32. 34. 36. contain the leading notes of their respective minor-keys, and have to be worked of course, in minor.

The Cantus firmus in the Middle Parts.

§ 3. When the cantus firmus is placed in the alto and tenor, the task becomes considerably more difficult, than when the former lies in extreme parts. Though the progression of the middle parts has to be a melodious, and independent one; still, when the cantus firmus is placed in one of the middle parts, it will be forced by its hemmed in position, to a more quiet and confined progression and cannot obtain that free, melodious formation, which soprano and bass are able to receive. Therefore, when the principal melody, the cantus firmus, lies in one of the middle parts, we have to consider before all, the soprano. This highest part may never adopt the quiet, confined character of a middle part. (Compare Manual of Harmony § 61.) The treatment of the cantus firmus, placed in the alto in the following manner in No. 37, would be most clumsy.

The above exercise would be somewhat improved, by changing the tenor with the soprano; we reproduce it in this form in No. 38.

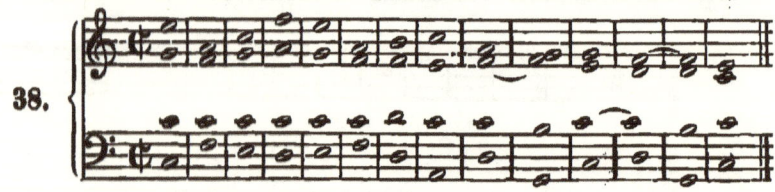

We add two more workings out of the same cantus firmus in the alto, in which the tenor progresses more melodiously than in No. 38, where it is only a replacement of the soprano, intentionally formed clumsily.

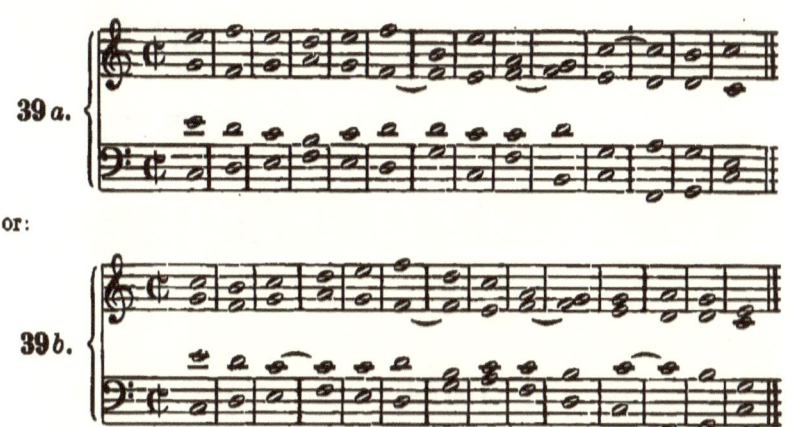

39 a.

or:

39 b.

We give now the exercises for the treatment of the cantus firmus, in alto and tenor. It is not advisable to stay too long with these exercises. The student can only obtain complete sureness in four-part writing later on, when he has mastered more complicated contrapuntal problems.

Exercises.

Cantus firmus in the Alto.

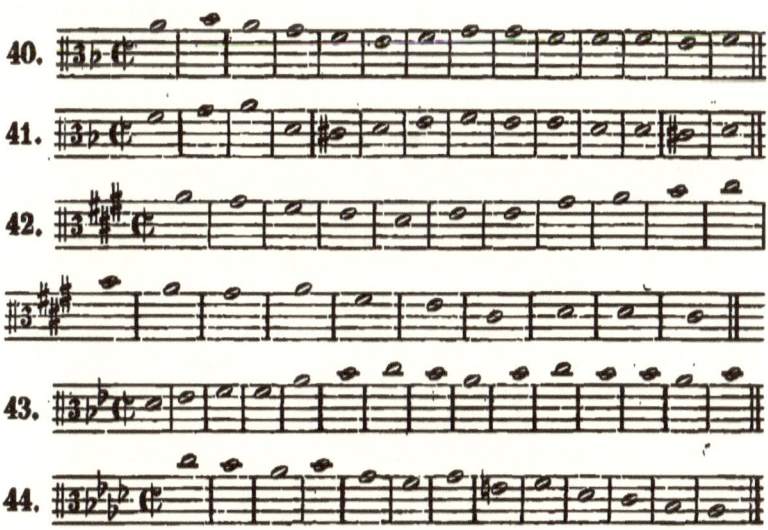

40.

41.

42.

43.

44.

Cantus firmus in the Tenor

CHAPTER II.

Unequal Counterpoint.

§ 4. In equal counterpoint, the parts can only progress inde-pendently, with respect to melody; in unequal counterpoint, however, the independance of parts is considerably heightened, by the freer rhythmical movement of one, or more parts, against the cantus firmus. Formerly one allowed 2, 3, 4, 6, even 8 notes to one of the cantus firmus, and practised this — at the beginning — by giving figuration to one part only. Here, it will suffice, if the student learns to place at the commencement, two notes, and later on, four, against one of the cantus firmus. All other species of movement — two or three divi-sioned time — will have to be reduced to the two above mentioned ones. On the other hand, we adhere to the procedure in giving the movement at first, to one part alone, although in practice it is more generally the case, that more than one part employs movement in turn, or simultaneously. Though it may prove more difficult to produce movement in one part only, still, just through this medium the attention is bent upon the proper progression of the individual parts. After the student has had sufficient practice in the manage-ment of each part alone, it will be an easy matter for him to work with freedom and sureness a partly simultaneous, partly alternative movement, between the different parts.

We commence now our studies by giving two notes to the bass, against one of the cantus firmus. *Each note of the counterpoint has to be a purely harmonious one.* In rare cases only, a suspension, well prepared by a leap, may be employed. This may take place either at the beginning, or shortly before the end, of the exercise, for instance:

Commencement.   Close.

49.

In the middle of a movement, the introduction of a suspension, even if well prepared, would make a disagreeable impression, as it leads to disturb the movement of the bass. That a suspension in the bass can only, in most cases, be employed before the third of a chord, has been shown in the "Manual of Harmony".

If then, a suspension, as disturbing the movement in the bass has to be used with care only, it stands to reason, that the tying of a note of a chord, to the same note of another harmony, has to be avoided altogether. There will then remain only the three following ways for movement.

1. A Leap from one note to another, of the same chord.
2. The Passing Seventh, from the octave of all chords of the seventh, dissolving downwards.
3. The fundamental note of a chord of the seventh, resounding from the chord of the sixth of a triad.

These three methods we see employed in the following three bars, namely: method first, in the first bar, method second, in the second, method third, in the third bar.

50.

The last bar of No. 50 b shows, that we may leave away the third of a chord on the second half of a bar; but it may never be omitted on the first part of it, and only very exceptionally, in a chord of the seventh.

In a few cases, the fundamental note of a chord of the seventh can strike after, provided, that retarded parallel-octaves are not evaded

by so doing. Ex. 51ᵃ cannot be found fault with; No. 51ᵇ is
quite inadmissable.

The open consecutive fifths and octaves, resulting from the want
of movement in the bass, will not be undone by the conspicious leap
of a sixth. A counterpoint, as the one here in No. 52, would be
entirely unallowable.

Still, in a few instances, the parallel-octaves are nevertheless
suspended by the movement. This is the case, when the bass, being
in the position of the chord of the sixth, resounds the fundamental
note of the chord of the seventh. Moreover, if contrary motion is
employed, especially in connection with two chords of the seventh
(53ᵃ), the effect would be a good one.

§ 5. It is not good to give to a counterpoint more than three notes of the same harmony, and, in the same direction. Consequently, the progression of the bass in Ex. 54 is bad.

The bass strikes here in the same direction; the notes *C, E, G, B* (chord of the seventh on the first degree of *C*-major;) after that, *C, A, F, D* (II$_7$) *A, F, D, B* (VII$^0_7$) and *F, D, B, G* (V$_7$). Such progressions should always be avoided most carefully. Towards the end, (last bar but one), the bass may very well make a leap of an octave, best however from below, but also from the higher to the lower octave. A leap of an octave — best from below to the higher octave — can also be employed advantagiously, at the beginning of the exercise (first bar). In the middle of a movement, progressions of octaves should only be used exceptionally.

§ 6. The last bar seldom, if ever, contains movement, also the first bar can do without it; occasionally, the bass may commence on the second beat on the bar.

or:

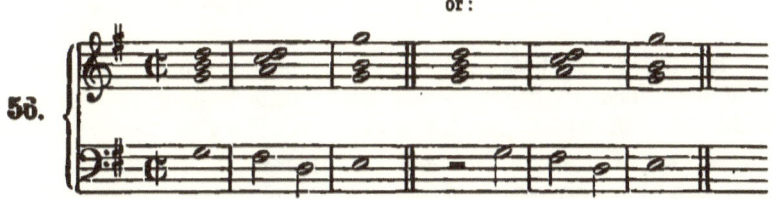

Any other passing note, except that of the seventh, descending from the octave, is not at all admissable. The progression of the bass in No. 57^a is bad, but the one at 57^b is good.

a. Bad.   *b.* Good.

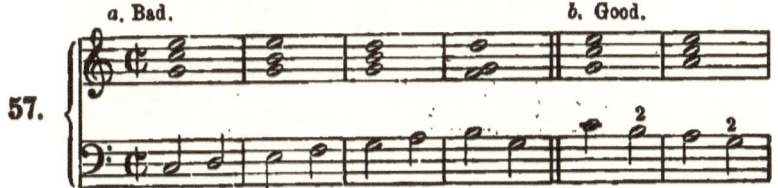

As the passing seventh allows the bass, only diatonic progression, and, considering that this part must mostly move in leaps, one can exceptionally, or occasionally, give two chords to one note of the cantus firmus, provided that, if by so doing, it affords the bass an opportunity of moving diatonically, granted that the progression of harmony be clear, natural, and comprehensible, as demonstrated in the following Ex. 58 NB.

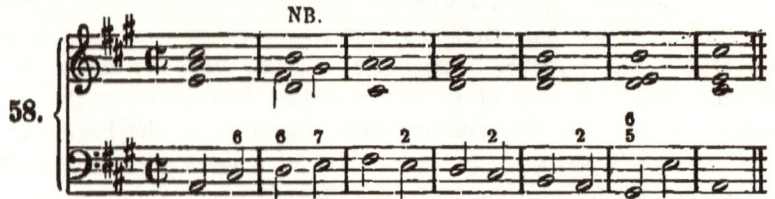

58.

§ 7. The passing seventh — from the octave of the fundamental note of the chord on the first degree in minor, — will have to be always the seventh degree of the *descending* scale in minor, which is not raised.

59.

In Ex. 59, the bass cannot progress in any other way than: *A*, *G*, *F* etc. The 7th degree of the minor scale is therefore not raised *in this case*. Retarded parallel-octaves, as shown in second beats in several bars of example 60, cannot be permitted, especially, as the equal and stiff progression of the extreme parts is not a good one.

60.

However, from the first note of an accentuated part of a bar to the second note of a non-accentuated one, the retarded parallel-octaves are completely covered by the two notes lying between, and are, therefore, allowable..

Single parallel-octaves, retarded by the movement from the second, to the second beat of a bar, can be allowed, but this must not be repeated through several bars, as demonstrated in Ex. 60. Example 62 is not to be censured, but Example 63, showing the same beginning, is bad, on account of its further progression.

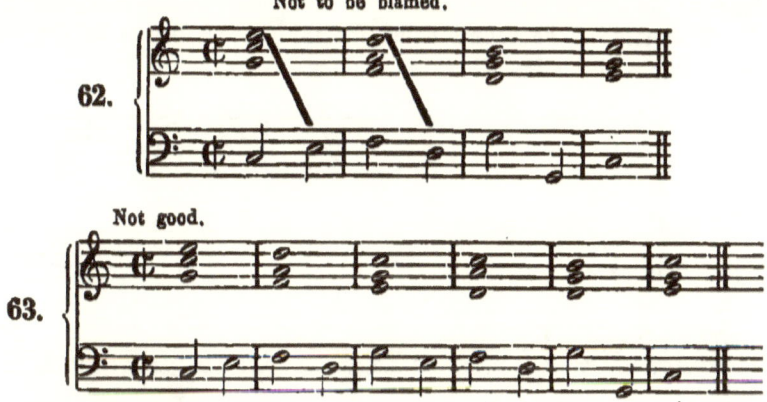

§ 8. We give now some exercises to the student of a counterpoint of two notes, against one of the cantus firmus, which latter can be used for alto, as well as for tenor.

The Cantus firmus is in the Soprano.

At NB., the seventh *F.* in the alto has to proceed upwards, on account of the passing seventh *F.* in the bass, which must descend. Likewise at NB.[2], the seventh *F.* in the alto has to proceed upwards, as the bass leaps into *E.* the original note of resolution. (See Manual of Harmony § 43.)

The retarded parallel-octaves, between the extreme parts NB. which fall on the non-accentuated beat, are admissable. (See Example 62.)

Now follows the counterpoint for the cantus firmus in the alto; see Example No. 44.

This example requires no further explanation. We give now a working out of the cantus firmus in the tenor under No. 48, with counterpoint of two notes against one of the cantus firmus.

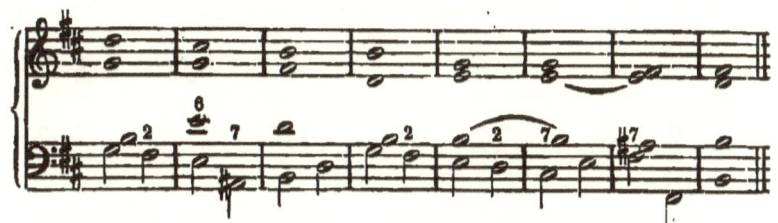

The beginning of example 65 might be also done as follows:

67.

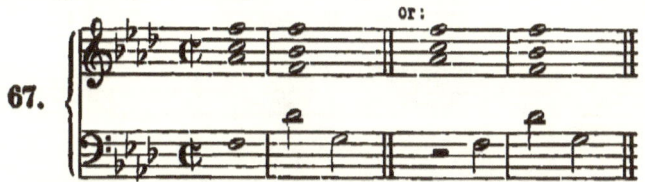

In order to give to the student as much practical guidance as possible, we still add a few workings of the following cantus firmus; only the last of them contains a modulatory deviation.

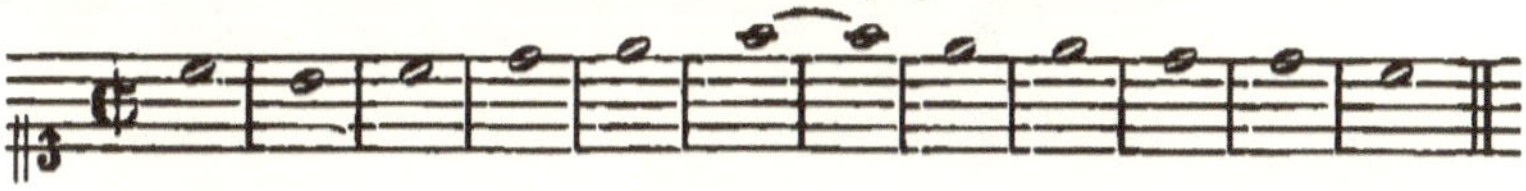

68.

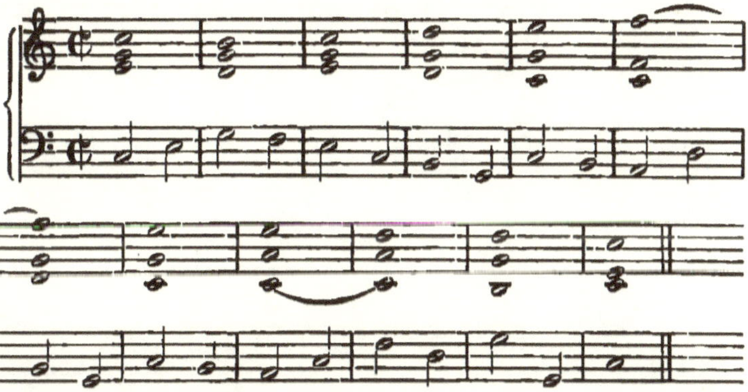

69.

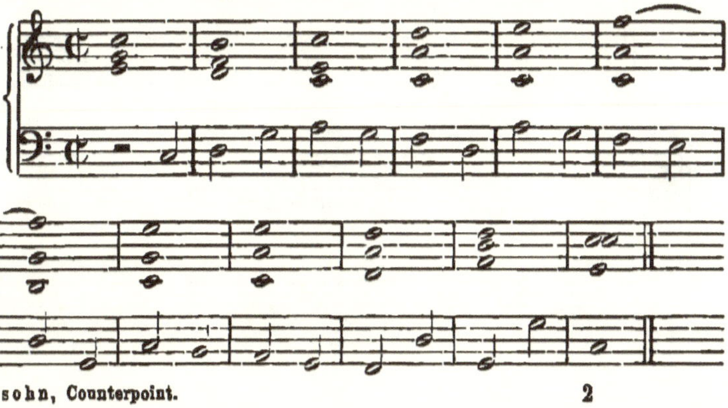

73.

One chromatic passing note, as at NB. in Ex. 73, can be used occasionally; such a note takes then the character of an altered tone, ascending from the natural tone. But this does not give an actual modulatory effect. In the following manner, the commencement of the foregoing cantus firmus can be treated:

74.

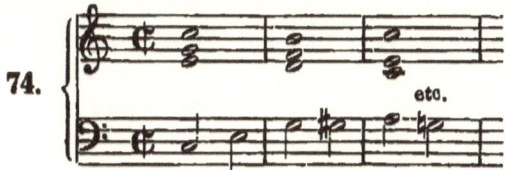

Several of such chromatic progressions however, should not be employed one after the other, as in this case the progressions of the parts would become, what old writers used to term: a *"howling progression"*. The nature of true Counterpoint is of a diatonic-melodic character. The following exercises would therefore be entirely objectionable. (Compare Manual of Harmony § 57. Ex.: 301 and 302.)

75.

The passing seventh in the alto, last bar but one, is, at the close, always allowable in all parts.

§ 9. The student may work now a few exercises, by placing two notes in the bass, against one of the cantus firmus. He may choose for the soprano one or the other cantus firmus, from exercises 29—36, and for alto and tenor 40—48. In the treatment of these exercises in unequal counterpoint, the pupil should not try to find support in the exercises he has already done in equal counterpoint; or, to bind himself to the employment of formerly used harmonies. He would not find his task easier, but more difficult. If he kept always to the same harmonic treatment, the mechanical patch-work, in putting a second half note on the other half bar of the bass, would be most inartistic. One would also soon make the observation, that what was good and suitable for work in *equal* counterpoint, would be often inapt for treatment in *unequal*. In attentively noting the examples No. 64, 65, 66, 68, 69, 70, 71, 72, 73, it will not escape observation, that in the counterpoint of the bass, the fifth of a triad, striking after, on the second beat of the bar, has been used but seldom. (Examples 65 and 71.) Now it is not in any way forbidden to allow the fifth of a triad to strike after; the succeeding counterpoint cannot be censured, although it shows a resounding fifth of the chord, of the first and fourth degree.

76.

Such resounding fifths are therefore not exactly forbidden, either at the triad, or at the chord of the seventh, where they form, on the second beat of the bar, the chord of the third and fourth; — but one cannot lose sight of the fact that, the frequent use of the fifth, striking after the common chord, will give to the counterpoint a weak, lame, and clumsy character. We warn therefore, against too frequent employment of this progression. For this reason Example 77 would not be recommendable, although it does not violate any of the stipulated rules.

77.

Those resounding fifths of the common chords, marked with * are easily avoidable, as shown in Example 78.

At the end of this chapter, we wish to draw attention to the fact, that it is not advisable to keep the pupil too long at these, not at all easy, exercises. In practice, mostly mixed counterpoint is employed. Besides, similar exercises are repeated in two- and three-part counterpoint. — Also in instrumental or vocal studies, one would not detain the student at the same exercises, until he has mastered them to perfection; by progressing to other, new studies, he will learn to overcome by degree the preceding difficulties with much better ability. As soon as the pupil has attained *some* efficiency in the formation of counterpoint with two notes in the bass, it will be advisable to proceed to the next chapter.

CHAPTER III.

Counterpoint in the upper parts.

§ 10.　For the treatment of counterpoint in the upper parts of two notes against one of the cantus firmus, there are eight methods available; these are:

1. The Leap of one note to another of the same chord.
2. The Suspension.
3. The Tie of one note of a foregoing harmony to the same note of another harmony in the next bar.
4. All Passing sevenths, which descend diatonically.
5. A Leap into the fundamental note of a chord of the seventh, to which the notes of the triad have been struck on the accentuated part of the bar.
6. The Leap into the Dominant Seventh, as also into all Minor, and Diminished Sevenths, by which the two latter will be enabled to serve as preparations for a suspension. A Leap into a Major Seventh has always to be avoided; a suspension therefore, by means of a major seventh, can be only ex-

ceptionally effected under certain conditions, viz. in a sequence of suspensions; (see example 85b. bar 3).

7. The Suspension before the fundamental note of the Principal Common Chords of the key, notwithstanding this note being already in one of the middle parts. But it is necessary that the fundamental note be under, and distant a ninth, from the suspension.

8. The Suspension of the Passing Dominant Seventh, prepared diatonically by the octave, when this suspension is placed between two or more suspensions or bindings, and if by this, to a certain extent a sequence is caused.

Note. The reason why the seventh cannot be used for the preparations of suspensions, except in those cases, mentioned under No. 6 and 8, is easily recognizable. The sevenths are dissonances themselves, and as such need resolving. Only, the leap into the minor and diminished sevenths, gives strength and power of resistance to these intervals, to support and carry the succeeding dissonance. Those instances, mentioned under No. 6 and 8, are explained by the exceptional character of the Sequence.

Examples to these eight rules:

1. The Leap.

2. The Suspension.

3. The Tie.

4. The Passing Seventh.

One can also write Sequences without hesitation, as noted at 82 b.

The fifth, striking after the chord of the sixth, is explained as the seventh of an imperfect chord of the fifth and sixth, on account of the fundamental note of the primary chord of the seventh, having been heard just before in the same part.

5. The Fundamental Note of chords of the Seventh, striking after.

This method will be used but seldom, and mostly in such a manner, as demonstrated in the four bars of No. 83. The chord of the seventh must appear in this case complete in all its parts as chord of the fifth and sixth. Progressions, as those, shown under No. 84, are not recommendable, although they may be used sometimes in practice.

In example 84 a. the chord of the fifth and sixth sounds empty; the third is wanting, which is the fifth of the fundamental chord. At b. the sixth is missing to the chord of the third, fourth and sixth, which is the third of the fundamental chord. At c. the Fundamental Note, of the chord of the seventh on the first degree, appears diatonically and not, as it ought to be, by a leap, therefore the effect is weak, although the chord of the fifth and sixth appears complete with all its intervals on the second half of the bar.

6. The Leap into the Minor and Diminished Seventh, for the purpose of Preparation for the Suspension, and the Suspension, prepared by the Major Seventh in a Sequence.

7. The Suspension before the Fundamental Note of the principal Triads, although the dissolved tone of the suspension be present in one of the middle parts. (See for reason and examples "Manual of Harmony" § 53. Exercise 257 b. c. and d).

Suspension before the Fundamental Note of the Triad of the Tonic.

Suspension before the Fundamental Note of the Triad of the Dominant.

Suspension before the Fundamental Note of the Triad of the Sub-Dominant.

8. The Suspension prepared by the diatonically Passing Seventh of the Dominant.

An instance showing this, is demonstrated at NB. of the example No. 87. bar 5th 6th 8th 11th.

§ 11. We commence now our work first with the counterpoint in the soprano on a cantus firmus in the bass, and show the employment of all the eight methods in one example. For better comprehension, we mark every first employment of each method with the corresponding number.

The student need not imagine, that he is obliged to make use
of all the eight methods in each individual exercise. On the contrary,
he is strongly advised to employ only the most usual ways, which
are those mentioned under 2. 1. 4. 6. 3. We note them down in
the succession in which we consider them mostly adapted for their
more frequent or rarer use. In accordance with that, the Suspension
would be the most suitable means, the Binding (especially, when
used several times through several bars) — the one least adapted,
for the movement of counterpoint in two notes. More than two successive leaps are not in accordance with the diatonic-melodic
character of counterpoint. The succeeding counterpoint would not be
advisable on that account, although it does not violate any of the
established rules.

Not good, on account of too many leaps.

As a rule (in working these exercises) one will do right, not
to confine oneself to one, but to change with the most usual means,
as far, as their employment seems to be adequate to the want;
and to make use of the less customary methods, (those mentioned
under 5. 7. 8.) only, when the progression of the counterpoint seems
especially adapted for their employment. Only the Suspension may
be used through several bars in succession; however one should not
capriciously amass them too frequently. It will be left to the good
taste and musical training, to decide in each individual case, which
method should be employed for the movement of the counterpoint.

The best proof of the excellency of a counterpoint will be always
its adaptability for singing; of course a sound, and natural harmonious
connection is self-understood. Sequences in the counterpoint should
not be used oftener than three times in succession.

The first four, even six bars of exercise 91 cannot be found
fault with, the continuance however, in a similar manner, produces
monotony. The use of Sequences at the beginning of a movement,
should not be repeated oftener than three times. In this respect the
commencement of exercise 92 might be called good, as were also
the first bars of exercise 91.

If the cantus firmus itself shows progressions of a decided se-
quencial character in the form of a cadence, it will be advisable, to
take also the other parts of the counterpoint in a sequence, viz:

or :

§ 12. Here follow some examples of a counterpoint in the soprano to a cantus firmus in alto and tenor.

Cantus firmus in Alto.

Cantus firmus in Tenor.

The student may now place a counterpoint of two notes in the soprano against the cantus firmus alternately in the bass, alto, and tenor of the following examples. As required, the cantus firmus of some preceding exercises may be treated over again for counterpoint in soprano.

Exercises.

Cantus firmus in Alto.

Cantus firmus in Tenor.

§ 13. For the counterpoint in the alto or in the tenor, the above rules (§ 10.) will hold good. The seventh method can only be used in the alto. Example 111 shows this case twice in the bars marked by NB.

Cantus firmus in Soprano.   NB.   NB.

Cp.
111.

The movement in a middle part alone, is very much more difficult than in an extreme part; we shall be reduced here to the methods mentioned under 1. 2. 3. 4. viz. the leap, suspension, binding, and the passing seventh. It will be unavoidable, for the middle parts, not to transgress occasionally the space of an octave. This however should not last for long, as the exercise would otherwise sound empty. The following treatment of the cantus firmus of No. 111. would therefore be worthless.

Bad, on account of too great a distance permanently between the middle parts.

Here follows for the direction of the pupil the working-out of a cantus firmus in a middle part, in six different examples.

Cantus firmus in Soprano, Counterpoint in Alto.

Cantus firmus in Soprano, Counterpoint in Tenor.

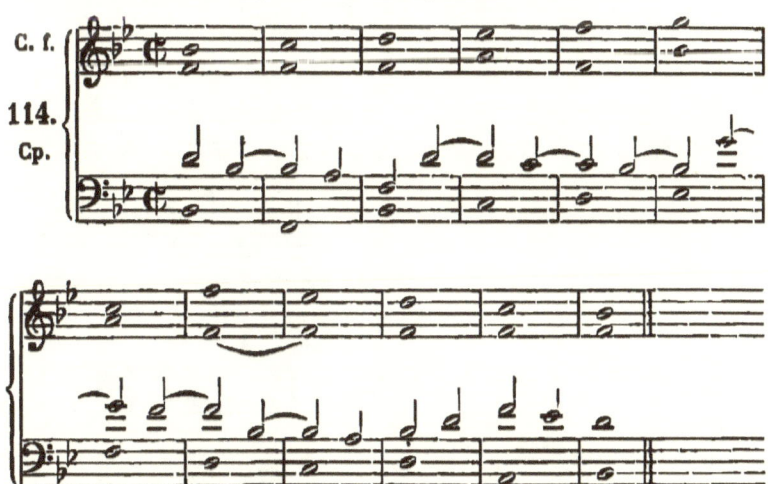

Cantus firmus in Tenor, Counterpoint in Alto.

Cp.
115.
C. f.

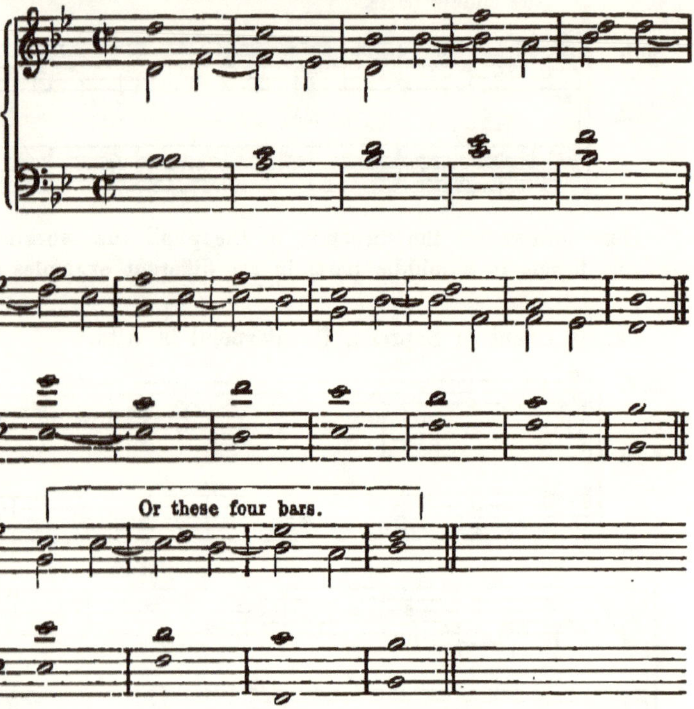
Or these four bars.

Cantus firmus in Alto, transposed to E♭-major, Counterpoint in the Tenor.

C. f.
116.
Cp.

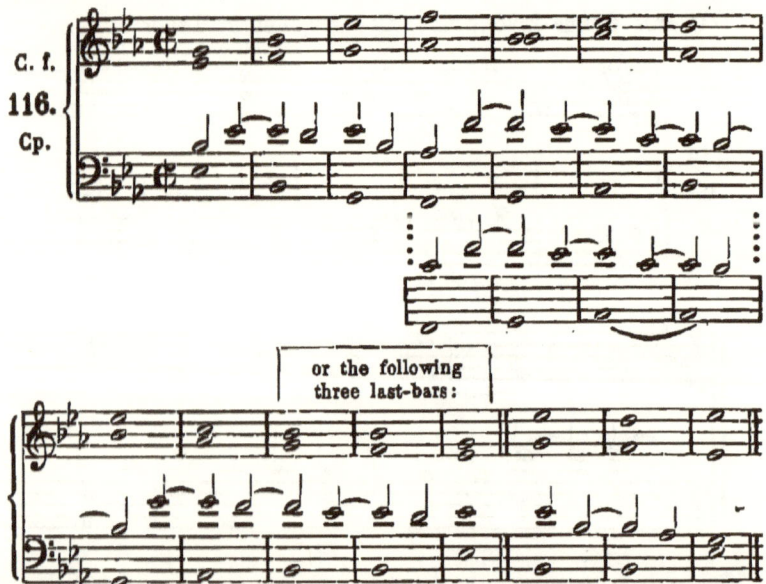

or the following
three last-bars:

Cantus firmus in Bass, transposed to G major, Counterpoint in Alto.

Cp.
117.

C. f.

Cantus firmus in the Bass, Counterpoint in Tenor.

118.
Cp.
C. f.

or the succeeding four bars
with a modulatory turning:

or also:

Exercises.

Note. For alto and bass the cantus firmus will have to be transposed to a convenient lower key.

119.

§ 14. After having progressed thus far, we arrive at working the counterpoint partly alternately, partly simultaneously in two or three parts, against the cantus firmus.

The purely mechanical method for this would be, to work the cantus firmus first in equal counterpoint, and then to add movement at proper places, to such part or parts, where it seems to be most adaptable. But we do not wish to recommend this manner in any way to the student. He must not put contrapuntal movement into a phrase, without having previously, paid proper attention to the movement of the parts. One may allow such a way of treatment to the uninitiated beginner for his first attempts; soon, however, he should accustom himself to conceive the composition in a freer and more artistic manner, which invents and considers the movement of the cantus firmus at the outset in close connection with the parts to one another. The student, however, must guard against overloading his work continually* with two or more contrapuntal parts. The simultaneous movement of two or three parts can be *occasionally* of very good effect; too much movement in several parts, at the same time though, fatigues. There follow now eight workings-out of the cantus firmus for the guidance of the pupil. No. 120. The cantus firmus has to serve for two examples alternately in different parts. The first example is meant to be worked always simply and in such a manner, that only *one* part takes the movement alternately with another part; the second, in return, richer in movement, and between two, even three parts alternately.

The Cantus firmus in Soprano.

The Cantus firmus in Bass.

124.

The Cantus firmus in Tenor.

The Cantus firmus in Alto transposed to E♭.

128.

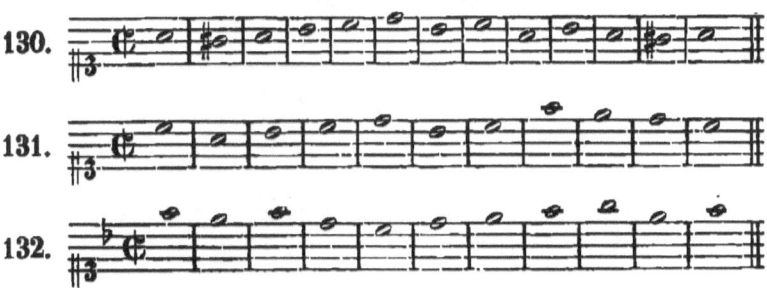

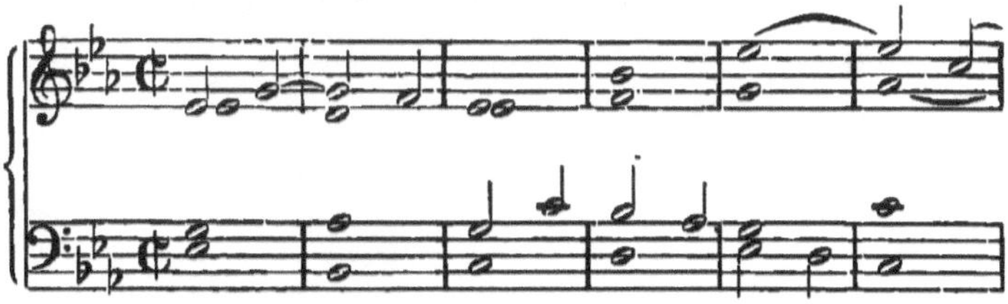

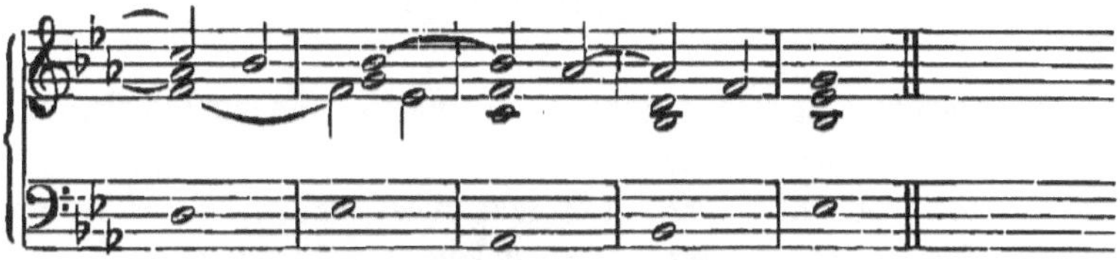

129.

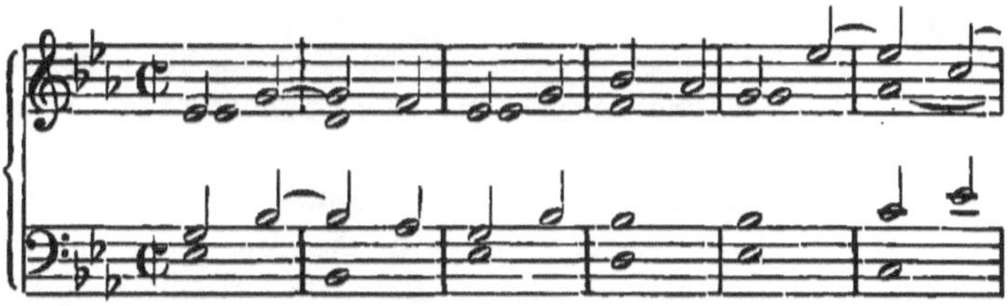

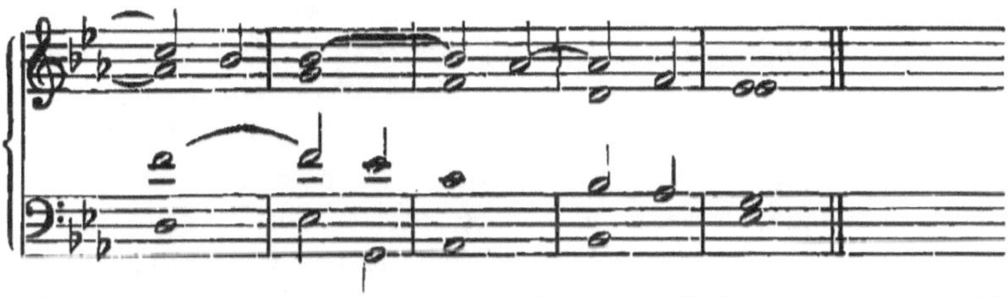

The following exercises are to be treated in accordance with the manner, shown in the examples No. 122—129. The cantus firmus may be transposed into other keys for bass or alto, according to the position of the part.

Exercises.

130.

131.

132.

If considered necessary, some suitable cantus firmus from former exercises may be chosen and employed in the manner, indicated in the examples 122—129.

CHAPTER IV.

Counterpoint of four notes against one of the cantus firmus.

§ 15. For the movement of four notes against one of the cantus firmus, (four crotchets against a semi-breve,) the same rules apply for *all* parts.

1. The first note of each bar must be an harmonic one.

2. Between two harmonious notes, passing ones may be inserted diatonically.

3. Changing notes are to be avoided; in the beginning of a bar they would be incompatible with the first rule, in the middle, opposed to the second rule. But we will not exclude them altogether from contrapuntal work. They will find their place at the more complicated exercises of the canon and fugue and can produce sometimes, even, a very excellent effect. Ordinarily speaking, one will do right to avoid them if possible, in all contrapuntal work, even in the canon and fugue; as the note of change, no matter, whether introduced from above or below, will always have the character of an ill-prepared Suspension, and is therefore not suitable for really "pure harmonic-structure". (Compare Manual of Harmony § 57).

The movement of four notes in counterpoint does not allow at all of harmonic binding, the Suspension but seldom, and exceptionally; at all events, its introduction requires to be done by leap. The following preparations will therefore not suffice.

133.

All such preparations of suspension are bad.   Worst of all is
the one at *d*, as here the resolution of the passing seventh (which is
moreover a major one) is retarded, in order to serve as a preparation
for a suspension.

Also the Suspension, prepared by leap, for instance:

134.

should be employed but seldom.

Its employment will be best suited for the end.   Many suspensions disturb and interrupt the flow of the movement.   Contrapuntal
progressions of this kind exhibit — so to say — a more modern
manner, than is usual in "serious style"; for instance:

135.

It is just the necessity of the preparation by leap, which in-
jures the diatonic melodious progression of parts.   The requirements
for really good counterpoint are always diatonic-melodious ones.   We
must therefore forbid, at a movement in crotchets, all figures of
chords, which do not contain, at least, one diatonic step (passing
seventh, perhaps also the ninth, striking after from the tenth).   This
diatonic step must, above this, be formed by the last two crotchets.

Bad.   Good.   Tolerably good.

136.

To be abstained from totally, are such figures, formed from the triad:

A mere circumscription of a semi-breve by four crotchets, through several bars, cannot be too carefully avoided.

Example 138 shows us such progression, which is totally inapt. Often, circumscriptions of a semi-breve cannot be avoided; but in such a case a change with the following figures will be advisable, inasmuch, as they can be used advantageously.

By a change of such figurations one would be enabled at a pinch, to circumscribe the following semi-breves.

140.

But even here we earnestly warn the student against this purely mechanical manner of working these exercises first in equal counterpoint, and to circumscribe then the semi-breve by four crotchets, which have to produce the movement. Such counterpoints show in most cases very distinctly their constrained origin. Such progression of parts and harmony, most suitable for *equal* counterpoint, does not always show the same adaption to serve an *unequal* one.

In the minor key the melodious scale is employed almost exclusively for diatonic progressions. The use of the augmented second of the harmonic minor scale must *always* be avoided. Descending, it may be used *sometimes* in very complicated problems (canon and fugue). But at our present exercises in single counterpoint, we will discard it altogether. In employing the melodious minor scale, we cannot be too careful, that the moving part does not strike a chromatically altered tone, while another part sustains the natural note, or vice versâ.

But even, if no other part sustains the natural note to the chromatic one, or the chromatic to the natural, we have to avoid those tones, which do not belong to the harmony. The following progressions are, therefore, impossible.

141.

C. f.

The F# is impossible in the second bar, because of the

third of the triad of the fourth degree being F; in the fourth bar, the soprano cannot strike G, while the tenor sustains G♯. One has therefore to manage the counterpoint differently; for instance:

142.

C. f.

We present now a few examples of a cantus firmus in minor. The counterpoint being alternately divided between the four parts.

Cantus firmus in the Bass, Counterpoint in Soprano:

143.

Cantus firmus in Soprano, Counterpoint in the Bass.

144.

Cantus firmus in Alto, Counterpoint in Tenor.

C. f.
145.
Cp.

Cantus firmus in Tenor, Counterpoint in Alto.

Cp.
146.
C. f.

Exercises.

147.
148.
149.
150.

The cantus firmus has to be transposed, when allotted to other parts.

§ 16. In the preceding examples, the pupil has bestowed his principal attention on the development of the one part, containing the counterpoint: he may now advance to the following studies, which occur often in reality. The counterpoint has now to be given alternately to the three parts in such a manner, that first one, and then the other, takes up and continues the movement. Also two, even three parts, may execute the movement. But this must not occur too often, otherwise the phrase would be overloaded by contrapuntal parts. No new rules are required for this treatment; we wish only to recall to the memory of the student (Manual of Harmony § 56 example 291) that at movement of the counterpoint in four notes, against one of the cantus firmus, the major and minor ninth may enter free by leap, when the latter is assisted and carried by the seventh. We show this in example 151.

151.

C. f.

The free entrance of the ninth, assisted by the seventh, is marked by an *.

As guidance for the treatment of the succeeding examples, we show here to the student a cantus firmus with movement in the three other parts.

Cantus firmus in Soprano.

C. f.

152.

Between suspension and resolution, notes may be inserted (Manual of Harmony § 56 Exercise 289.

Cantus firmus in Bass.

153.

C. f.

Cantus firmus in Alto.

C. f.

154.

Cantus firmus in Tenor.

155.
C. f.

Phrases of imitation, as contained in the last bars of example 155, render a peculiar charm to contrapuntal parts.

Exercises.

156.

157.

158.

159.

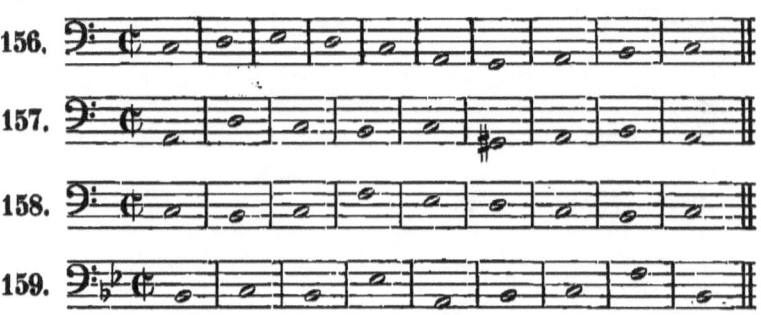

Remarks on these Exercises.

Modulations, and short evasions into nearly related keys, can be permitted occasionally. Now and then a chromatic passing note, or still better an altered fifth, may be used; for instance:

160.

C. f.

Several successive chromatic notes should, however, be carefully avoided. (Manual of Harmony § 57 Examples 301 and 302.)

CHAPTER V.

The Three-part Phrase.

§ 17.   The three-part phrase in equal counterpoint should be worked thus, that the harmony be clearly recognizable, although there are only three parts available for the representation of four-part chords.   This can be easily effected, as the middle part, (no matter, whether tenor or alto,) affords more room for independent progression. The middle part will, therefore, be allowed to move oftener in leaps, especially in fourths and fifths, than in four-part writing; and this is advisable, when, by so doing, the harmony can be made fuller.   The distance of the alto from the soprano may amount to a tenth, nay, even occasionally to an eleventh.   Beginning and end will be best rendered in unisons.   The chord of the sixth, on the seventh degree, would sometimes take the place of the chord of the seventh.   Similar motion of all the three parts will have to be avoided; but the chord of the sixth, on the seventh degree, may follow exceptionally the chord of the sixth, on the first degree, in a descending direction.

But not so good is the reverse.

But even this can be allowed occasionally.   Progressions though, such as follow, have to be always avoided, as they are diametrical to the nature of counterpoint.

The chord of the seventh may be used sometimes without a third. (Compare Manual of Harmony § 36, note). Hidden octaves cannot be avoided at the close, when the three parts finish in unisons. Moreover, the student may be reminded that all kinds of hidden fifths or octaves will be much more noticeable in three-part, than in four-part writing. The treatment of the parts requires therefore more care. We will endeavour to illustrate, in the three following examples, the treatment of a cantus firmus in equal counterpoint with three parts.

Cantus firmus in Soprano.

Cantus firmus in Bass.

In the last bar but one, the stationary bass, which contains the cantus firmus, is made good by the decided movement of the upper parts.

Cantus firmus in Alto.

If we wish to write to this cantus firmus a contrapuntal movement in minims in one or the other part, alternately or simultaneously, we shall have to comply with the same rules, as given for
the four-part phrase. It will be more advisable at the three-
part phrase, to use also, at the movement of the counterpoint in
minims, occasionally a passing chromatic note, when by so doing the
harmony gains in fullness. From this point of view, the G♯ on
the second half of the fifth bar in example 165, and the A natural
in the third bar in 167, cannot be censured. We directed the bass
purposely in this manner, to show the student, that he may use occasionally, but not too frequently in succession, such progressions. It
scarcely needs mentioning, that in both cases, the use of the chromatic
note could have been easily avoided, as is shown at the close of the
corresponding examples.

Cantus firmus in Soprano.

At B. the proper tone of resolution of the suspension- D, has been left away, in order to make the harmony fuller (Compare "Manual of Harmony" § 56).

The treatment must be the same, if the cantus firmus is situated in alto or bass. The student may practise this, with this cantus firmus, as also with any of the canti firmi employed further on.

§ 18.   The movement of four notes in the counterpoint, against one of the cantus firmus, will have to be considered also by the rules, given for the four-part phrase. Here follow three examples relating to this; the cantus firmus is in the alto.

NB. Old authors used to employ the full tone G; The use of the semitone
G♯ is modern.

The treatment has to be continued in the same manner, when
the cantus firmus is placed in the soprano or bass; the cantus firmus
of the succeeding exercises has to be employed in equal counterpoint,
and also with movement in all the parts, as demonstrated in examples
161—169.

Exercises.

174.

175.

CHAPTER VI.

Counterpoint in Two Parts.

§ 19. When we have to form a phrase by two parts in equal counterpoint, we have to commence in unisons or on the octave, sometimes also with the perfect fifth, and to close the phrase by means of unisons or octaves. *In the middle of the phrase no perfect consonance should be found. Unisons, octaves, perfect fifths, and fourths are therfore to be excluded.* We can only employ Imperfect Consonances: the major- and minor-thirds and sixths; and dissonances, the augmented fourth and diminished fifth. These intervals are the most suitable for making the harmonies in two-part writing most concise and recognizable. The minor seventh and major second are not uitable for this p hrase. Successions of thirds or sixths, through more than two, or the utmost, three bars, are to be avoided, as contradictional to the character of counterpoint. It is not wise to remove the two parts further from one another, than a tenth; in unequal counterpoint, however, an occasional transgression of this distance to the twelfth, may be permitted. and the two parts may only be removed so far, transitorily, as their mutual assistance would not be effective. All hidden fifths and octaves have to be avoided; even the hidden octave over the semi-tone and the leading note are interdicted. (Manual of Harmony § 59). One cannot therefore write thus:

The close will have to be formed by contrary motion; viz:

We give now an example in equal counterpoint:

176.
C. f.

We give the succeeding rules for the unequal counterpoint in semi-tones:

1. One can make use of the Suspensions of the Fourth before the Third, the Fifth before the Augmented Fourth, and the Seventh before the Sixth.

2. The Perfect and Diminished Fifths may be used diatonically on the second beat of the bar, after the sixth. This passing fifth takes then the character of a passing seventh, and has to descend diatonically; viz:

177.

3. The Fourth striking after by leap has to be avoided.

178.

In return the diatonically descending fourth is allowable, as it bears the character of a passing seventh and is dissolved downwards for instance:

179.

4. The Fourths, Fifths striking after, are permitted.

180.

5. A Succession of Two Major Thirds are prohibited.

6. The Passing Seventh is allowable at the beginning of the phrase, for instance:

181.

Here follows an example of this kind; it contains a counterpoint in soprano, and one as well in the tenor. Both treat the same cantus firmus.

182.

Concerning the counterpoint of four notes, against one of the cantus firmus, the movement may be commenced sometimes, with a third or a fifth; viz:

183.

As for the rest, all rules already given, apply also here; we show such a treatment under No. 184.

184.

We subjoin here the following remark. It is not very probable, that in our time, accustomed to the use of full and rich harmony, a lengthy and elaborate contrapuntal movement will ever be written for two *singing* voices. In return, a short Intermezzo in a vocal composition of more voices, (for instance in a vocal fugue) might be often of very good effect. It would serve, then, as a contrast to preceding richer polyphonic formations and öffers to the ear, so to speak, a resting point. We have to look entirely differently upon a two-part instrumental movement. By means of much elaborate figuration, the harmony can be indicated, and will appear consequently, fuller and more complete. We can notice this for instance in two-part fugues and refer to SEB. BACH's E minor fugue, No. 10. volume I of the »Wohltemperirte Clavier«. •

Exercises.

PART SECOND.

Double Counterpoint.

CHAPTER VII.

§ 20. We call a counterpoint double, when it is formed in such a manner, as to allow it's removal an octave, or tenth, or eleventh, against the part, that it has been worked. We consider only three kinds of double counterpoint: the one in the octave, the tenth and twelfth. Older treatises contain also rules for the double counterpoint in the ninth, eleventh, thirteenth aud fourteenth. Such counterpoints, suitable to be inverted into so many different intervals, can only be formed under such very limited conditions, that they will find very rarely employment, perhaps never, in real practice. We commence with counterpoint in the octave in two-part phrase. We need only add to the rules, already known for the treatment of the two-part phrase, that either parts should not be removed from one another more than an octave, on account of the inversion being in that case illusionary. An inversion of parts into the double octave would separate both parts too remotely from one another. We demonstrated an example under No. 188 of a double counterpoint, which could be placed to the cantus firmus in alto, as a soprano, and also as a lower part. We will not give in future the cantus firmus only in notes of equal value, as has been done hitherto in single counterpoint. By so doing the cantus firmus will become less rigid. At the formation of the counterpoint, however, we shall have to bestow especial attention, that the former is as much as possible contrasted in rhythmical respect with the cantus firmus.

At this kind of cantus firmus, a binding is allowed between two crotchets of one harmony to another. The crotchet, entering diatonically, may also serve as a suspension. This will be all the more

justified, as, in most cases, not four, but only two notes (two
crotchets) are placed against a minim of the cantus firmus.

One will perceive, that only those intervals and progressions are
possible, which were available in two-part writing in single counterpoint. Intervals, as for instance, the augmented sixth, should not
be used there, and must be excluded also here. We give under
No. 189 another example of this kind of counterpoint. The student
will perceive, that also modulations, which do not lead too far from
the principle key, will be serviceable.

Exercises.

190.

To this cantus firmus the lower part has to be invented; at the inversion the cantus firmus is placed into the lower octave, the counterpoint remains.

191.

To this alto the soprano has to be placed as counterpoint, and to be removed to the lower octave: the cantus firmus remains; similar in 192 and 193.

192.

193.

Double counterpoint in the octave, in the three-part phrase.

§ 21. When we place a soprano and alto to a cantus firmus in such a manner that, the soprano removed an octave lower, can be used as tenor to the bass and alto, we must adhere to the following conditions:

1. Soprano and alto must not be separated from one another more than an octave.

2. They must not move in such parallels of fourths, which would form, when inverted, parallels of fifths: viz:

194.

Parallels of fourths of this description: will be allowable; when the bass moves in contrary motion; viz:

195.

3. The soprano cannot be led nearer to the bass, than at a distance of an octave, as it would, when inverted, be placed underneath the bass.

196.

4. Likewise the real suspension of 9 — 8 ought to be avoided, as the result would be, when inverted, two before one.

197.

At a double distance of the soprano from the bass, this suspension is however avoidable, as, in this case, it remains still a suspension, nine before eight, after the inversion into the lower octave.

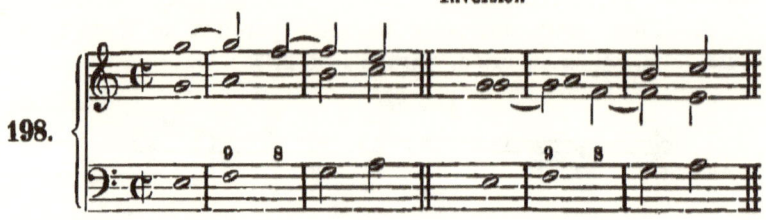

198.

Proper attention must be bestowed on the formation of the alto, inasmuch as the latter will become the upper part, after the inversion of the soprano into the lower octave.

Here follows an example for the demonstration of this kind of counterpoint.

Cp.

199.

C. f.

Inversion

C. f.

It will evidently remain exactly the same, whether tenor and alto are placed first to the cantus firmus in the bass, and the tenor appears then as soprano at the inversion into the higher octave. It is only necessary to pay the same consideration to the progression of the tenor with regard to its melodious formation, as was done in the first instance concerning the alto.

§ 22. If the alto has to be changed with the bass, the same rules have to be regarded, as those given for the inversion of soprano and alto; with the exception that the suspension 9—8 will have to be altogether excluded between bass and alto, and bass and soprano, as mistakes would always occur at the inversion, against the rules for the use of suspension.

One should therefore abstain from writing:

200.

C. f.

The Inversion would be;

Just as wrong would be the following suspension:

We alter therefore the counterpoint of the Alto in this manner:

We reproduce now the example under No. 201 with the cantus firmus in the Alto.

An inversion of the Bass and Soprano would be effected in the following manner:

First manner of inversion; the Bass is placed two octaves higher; Soprano and Alto remain.

Second manner of inversion; the Bass is placed two octaves higher, the Soprano an octave lower.

Third manner of inversion: the Bass one octave higher, the Soprano an octave lower.

Should we prefer to avoid the quickly passing crossing of parts in NB. in No. 205 (which bye the bye is quite without consideration) we would alter the alto, which is a free part.

In all these examples it is quite immaterial which part receives the cantus firmus originally. All exercises are treated by the same rules and principles, laid down above.

In the uneven time ³⁄₄ or ³⁄₂ the rules, given for the counter-point in even time, are in force. For the better comprehension of the pupil, we note an example in uneven time; he will learn from the preceding, as well as from the following examples, that the fifth of the triad is introduced in all places, where it is used, with great care, in such a manner, that there may not appear consequently, in one of the inversions, awkward sounding chords of the sixth and fourth. This shall also find proper consideration in these exercises.

First kind of inversion; The Soprano is lowered an octave, Bass and Alto remain.

The same kind of inversion will remain, if the alto be placed an octave higher, and the soprano remains as before.

Second kind of inversion. Bass and Alto are inverted an octave, the Soprano remains.

Third kind of inversion. The Bass is removed two octaves higher, the Soprano an Octave lower: the Alto remains.

210.

The Fourth kind of inversion we give now, (to avoid crossing of parts) by placing the Soprano two octaves lower, the Bass an octave higher the Alto remains.

211

Exercises.

Note. The student need not hesitate to transpose inversions into other keys, should he find the parts transgressing their individual compass.

CHAPTER VIII.

Double Counterpoint in the Octave in Four-part writing.

§ 23. The student can produce the simplest kind of Double-Counterpoint in four-part writing, by forming a movement in such a manner, that tenor and soprano can be exchanged, one with the other. There are no new rules required for this. Those given at § 21 and 22 will remain in force also here. Here follows an example in which tenor is placed in such a manner to the cantus firmus in the soprano, that both parts may exchange places. Alto and bass may be regarded as free parts.

In the inversion, the tenor takes the cantus firmus; the soprano gives the counterpoint of the tenor. Both parts are inverted into the octave.

Inversion of No. 216.

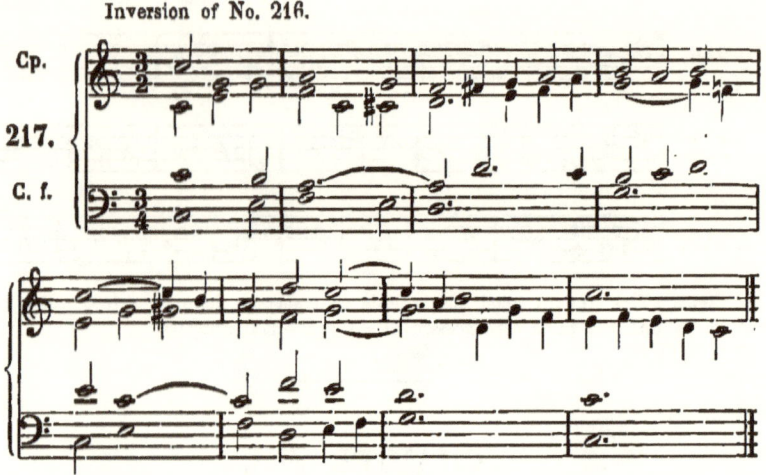

In example 216 however, the two parts, which are not meant to be inverted, are nevertheless also treated according to the rules of double counterpoint. We are able therefore also to exchange with each other, (besides those inversions already demonstrated), soprano and alto, alto and tenor, alto and bass, tenor and bass, and lastly bass and soprano, by which proceeding we gain still five more inversions of example 216. But we must here expressly remark that the pupil is not at all obliged to work his exercises *in this manner;* this would cause him a great deal of unnecessary difficulty. He is only required, at present, to contrive to work but one contrapuntal part to the cantus firmus; the other parts, which have not to be inverted, he may consider as free ones.

In placing then the cantus firmus alternately into another part, he must work several examples, in which he will have only to treat two parts in such a manner, that they can be inverted, without any regard to an inversion of the free parts.

The student may practise this problem in the order indicated at example 216. He will perceive by the inversions, that the fifth of a chord has to be introduced with especial care, when it appears in the bass in one of the inversions. In like manner the Suspension nine before eight is almost always unsuitable. Altogether the introduction and resolution of suspensions requires the greatest care. The employment of the Augmented Sixth will prove alike difficult. This interval will present itself, at the inversion, as a diminished third, and not only prove itself a harsh dissonance, but may also give rise to faulty progressions; for example:

Inversion.   Inversion.

The employment of the Augmented Sixth will have to be therefore avoided, in the chord of the fifth and sixth, as well as in the chord of the sixth. The altered fifth can occasionally be used, as is shown in example 216, bar fifth.

Here follow now the inversions of two other parts, which we will demonstrate in further inversions taken from the example 216.

The Cantus firmus in the Alto; the Soprano overtakes the counterpoint of the Alto. For melodious consideration this exercise has been transposed into F-major. Such transpositions into other keys, (which have been mentioned before,) are often necessary at certain inversions, when the phrase is meant to remain within the compass of the singing voices.

218.

The Cantus firmus in Soprano; the Alto replaces the Counterpoint of the Tenor, the Tenor that of the Alto. The inversion has been transposed into A-major for melodious considerations.

219.

The Cantus firmus in the Soprano; the Alto overtakes the counterpoint of the Bass, the Bass that of the Alto.

220.

The Cantus firmus in Soprano; the Tenor takes the counterpoint of the Bass, the Bass that of the Tenor.

221.

The Cantus firmus in the Bass; the Soprano replaces the counterpoint of the Bass.

222.

In the same manner other inversions can be worked, for instance: the change of the bass into the alto, the alto with the bass, (having the cantus firmus in tenor,) the bass with the soprano, cantus firmus in the alto or tenor, etc.

The pupil should work out the following exercise. To begin in as simple a manner as possible in equal counterpoint; having done this, he may proceed to finish the cantus firmus with figurated counterpoint, in the inverted and free parts. The practice of his own endeavours will prove clearly to him the necessity of the given rules, principles, and remarks on this kind of double counterpoint. The inversions of the exercises should be written down always, in order that experience may be gained of the real effect of the double formations of this kind of writing. They will often give rise to many corrections and alterations of the original work.

Above all, the student must pay attention to the various parts being formed independantly melodiously, and that the distance of the parts from one to another be sufficient, to allow for the inversion.

Exercises.

The cantus firmus may be given to every one of the parts. We demonstrate the manner of treatment with two free parts in the following examples: cantus firmus from No. 224.

Cantus firmus in Soprano, Counterpoint in Tenor; Alto and Bass are free parts.

223 a.

Inversion.

Cantus firmus in Alto, Counterpoint in Soprano; Tenor and Bass are free.

223 b.

Inversion.

Cantus firmus in Soprano, the Middle-parts are inverted.

223 c.

Cantus firmus in Soprano; Tenor and Bass inverted.

223 d.

Cantus firmus in Bass; Bass and Soprano inverted.

223 e.

Inversion.

Remarks on these Exercises.

It does not matter, if one or the other of the inversions commences or finishes with the chord of the sixth. The chord of fourths and sixths however has to be avoided at the beginning and close. Also in the middle of the movement attention must be paid, to what has been said, regarding the introduction of the fifth of a chord, which when inverted, would result in the chord of sixth and fourth. This chord cannot of course be avoided altogether. The student need not trouble himself too much to evade it; one has only to bestow sufficient care on its proper introduction, otherwise, this chord may easily sound weak or have a bad effect.

CHAPTER IX.

Triple counterpoint in the octave in three- and four-part phrase.

§ 24. If in a three-part phrase all the parts have been treated by the rules of double counterpoint, five Inversions can be formed from it, namely:

1. Position.	2. Position.	3. Position.	4. Position.	5. Position.	6. Position.
1. Soprano.	1. Soprano.	2. Alto.	2. Alto.	3. Tenor.	3. Tenor.
2. Alto.	2. Tenor.	1. Soprano.	3. Tenor.	1. Soprano.	2. Alto.
3. Tenor(orBass.)	3. Alto.	3. Tenor.	1. Soprano.	2. Alto.	1. Soprano.

In these Inversions one can, (to avoid disagreeable crossings,) place one or the other part two octaves higher or lower. In the ordinary inversion in the octave, it does not matter, when occasionally one of the lower parts crosses a higher one for a few notes; but this must not be continued through many bars, otherwise the effect of an Inversion would be lost. Below the bass, or the lowest part representing it, no other part should be placed even momentarily. At the beginning and close the fifth has to be avoided, in order that not one of the inversions commences, or closes the movement with the chord of the fourth and sixth. In the middle of the phrase, as well, all the rules, formerly given, regarding the position and introduction of the chord of the fourth and sixth, are to be observed. A Suspension nine before eight, has *always* to be avoided. All parts must be formed melodiously independently; as each of them inverted in their turn in the soprano, will become an upper part. For this reason the resting of one part, for any length of time, on the same note, would be impracticable, unless it be purposely meant, as in a Pedal-point. This however would not sound well in the three-part phrase. We give here an example of triple counterpoint with all the inversions. The Cantus firmus is placed at first in the Bass.

First inversion. The Alto is placed an octave lower and forms the Bass.

Second inversion. The Alto is placed in the higher octave and produces the Soprano. This and the next inversion, ought to be transposed to the key of *G.* to render them more practicable for singing voices.

Third inversion. The Soprano is placed an octave lower, Bass and Alto an octave higher; the Soprano now forms the Bass.

Fourth inversion. The Bass is placed an octave higher; the Alto forms the Bass in this inversion.

Fifth inversion. The Bass is placed an octave higher, the Soprano an octave lower. The Bass is Soprano, the Soprano Bass.

Exercises.

234.

235.

236.

After the student has worked out the cantus firmus of these exercises, he may try to invent, independently, similar phrases, which would allow five inversions.

If it be intended in four-part-writing to place to a cantus firmus the three upper parts in triple-counterpoint in the octave, the task would be in some respects rather easier than in the foregoing three-part exercises.

As none of the three upper parts have to form the bass in any of the inversions, we can introduce the fifths everywhere. Even the Suspension of the ninth before the octave can be brought into requisition, as we will show presently, in the subjoined example, bar five and seven. An occasional crossing of the parts, quickly passing in one of the five inversions, would not be of any moment, especially in the middle parts. Occasional transposition, will be found necessary,, when the exercises are meant for singing voices.

The Cantus firmus in the Bass.

First inversion (transposed to *C.*)

Second inversion.

Third inversion (transposed to *D.*)

Fourth inversion.

Fifth Inversion.

Exercises.

Quadruple Counterpoint in the Octave.

§ 25. In this kind of counterpoint the four parts will allow (excepting the original position) twenty-three inversions.

The phrase can therefore be represented in the following twenty-four positions. (For shortness we note the parts Soprano, Alto, Tenor, Bass by the numbers 1. 2. 3. 4,)

1 1 1 1 1 1	2 2 2 2 2 2	3 3 3 3 3 3	4 4 4 4 4 4
2 2 3 3 4 4	1 1 3 3 4 4	1 1 2 2 4 4	1 1 2 2 3 3
3 4 2 4 3 2	3 4 1 4 3 1	2 4 1 4 1 2	2 3 1 3 1 2
4 3 4 2 2 3	4 3 4 1 1 3	4 2 4 1 2 1	3 2 3 1 2 1

Also in this case we have not to mention any new rules; but the strictest observation of all the conditions of double counterpoint, already given for the relation of all four parts, will have to be adhered to, if the inversions have to prove usable.

In practice such movements with twenty-three inversions will be seldom required; one would certainly never produce all the inversions, (even if they should be perfectly usable,) within the compass of a piece of music. This would produce monotony. We advise therefore not to stay too long at this kind of counterpoint. In return we recommend urgently the practice in triple double counterpoint. Such movements occur very often in practice, as we will see later on in the treatise of the fugue.

Here follows now an example in quadruple counterpoint. Of the four and twenty displacements, we will only produce the four most important ones, to save space. Here are those inversions, in which each part changes its place.

$$
\begin{array}{cccc}
1 & 2 & 3 & 4 \\
2 & 1 & 4 & 3 \\
3 & 4 & 1 & 2 \\
4 & 3 & 2 & 1
\end{array}
$$

To write out the other inversions, will be left to the student; he will see that, provided the parts are worked properly by the rules for double counterpoint, the inversions would prove usable. Ex. 216 is also treated in quadruple counterpoint; besides those six inversions marked under 217—222, the student may also write out the other seventeen.

245 b.

245 c.

245 d.

245 e.

We show here also example 216 in the inversion $\frac{4}{3}\frac{2}{1}$

245 *f*.

Exercises.

246.

247.

248.

The student may also invent such movements, which are worked quadruple double counterpoint.

CHAPTER X.

Double Counterpoint in the Tenth and Twelfth.

§ 26. In double counterpoint in the tenth the question is, to invert a part, a tenth, or third. The intervals, which appear in the inversion, may also be noticed by the following table of numbers.

$$1 \quad 2 \quad 3 \quad 4 \quad 5 \quad 6 \quad 7 \quad 8 \quad 9 \quad 10$$
$$10 \quad 9 \quad 8 \quad 7 \quad 6 \quad 5 \quad 4 \quad 3 \quad 2 \quad 1$$

It is evident that at this kind of counterpoint the succession of two thirds, or tenths, or sixths, should not take place. They would result, at the inversion in parallel octaves, or unisons, or parallels of fifths.

Inversion.

The fourth and the seventh can only be used passing, in such a way, that the fourth is led into the fifth, the seventh into the eighth; for instance:

Inversion.

Inversion.

The suspension of the ninth is dissolved in this way:

Inversions.

It is clear, that at the double counterpoint in the tenth and eleventh, only contrary motion can serve the purpose, as those intervals generally used in parallel motion: thirds, sixths, and tenths have to be excluded here.

Older treatises put forward still a considerable number of rules concerning those intervals, suspensions and progressions, which were to be evaded or permitted. All those rules are certainly perfectly justified, but they serve in most cases only, as many years of experience has taught us, to embarrass the student. We, on the other hand, put forward but *one simple rule* for the double counterpoint in the tenth; and this contains all, which is required, for its formation.

Write the lower part to a higher one in this manner that one can add to it a higher third.

Or, write the upper part to a lower one in such a manner, that one can add to it the lower third.

Provided now the rules, relating to double counterpoint in the octave have been taken into account, one will be able to invert a phrase, treated in this manner, also in the tenth. Regard the subjoined example.

C. f.

249.

Cp.

We have placed to the cantus firmus of the upper part, the lower part thus, that we may add the lower third to the upper part, as well as to the lower part the upper third.   As the parts are treated, according to the laws of double counterpoint, in the octave, we shall now be enabled to invert each part a tenth or — which is equivalent — a third.   We show this in the following examples.

First kind of inversion.   The upper part is placed a tenth (or third) lower, the lower part remains.

250.

To this, and also to the succeeding inversions, one, or two free parts have, of course, to be added, as in practice such counterpoints, formed by two parts only, could not be used.

Second kind of inversion.   The upper part remains, the under part is placed a third (or tenth) higher.

251.

Third kind of inversion. Both parts are placed a third (or tenth) higher.

Fourth kind of inversion. The upper part is placed a tenth , lower, the lower one a tenth higher.

In a three-part phrase, we can treat this example in the following way, containing one part, worked in double counterpoint, in the octave.

We note only the commencements; the student may then, for practice, write out the whole example, as well as the inversions.

First manner.

253 *b*.

Second manner.

253 *c*.

It is also practical to add to the counterpoint, as a third part, another free one, which has not to be inverted; but then, it requires to be replaced anew to the two inverted parts, as middle and lower part, at each inversion.

In the same manner one can add two free parts to the two parts, containing the double counterpoint in the tenth, and construct the latter differently at each inversion.

We demonstrate this by furnishing example 250. (the first inversion of 249) with *one* free upper part, and example 253 a (fourth inversion) with *two* free parts (Soprano and Tenor).

253 *d*.

For the adaptation of singing voices, we transpose inversion 253a to *a*-minor.

253 *e*.

It is self-understood that this kind of double counterpoint can also be treated in four-part phrase by combining the above two kinds to represent it in three parts, for instance:

254.

We have notified this already in example 249 with small notes, and give now three inversions of this little phrase. It will be proved by this, that no other counterpoint exists *than that one in the octave.*

We add here still another example of the employment of such a double counterpoint. This is the "Stretto" (Engführung) in SEB. BACH'S fugue in *B* flat-minor (Wohltemperirtes Clavier Th. II.)

258.

The student may endeavour now to compose little movements in double counterpoint in the Tenth on his own account.

Double Counterpoint in the Twelfth.

§ 27. In comparing the following tables of numbers and notes, one will perceive that by inversion the Prime and Octave will be changed into Twelfths and Fifths, the Eleventh into the Second, the Tenth into the Third, and vice versa.

As the sixth becomes seventh by inversion, it must be always prepared, and descend one step diatonically; as passing note, when descending, it need not be prepared; for instance:

The double counterpoint in the twelfth is based essentially on the progression by thirds or twelfths. The monotonous combination requires to be cleverly concealed, by giving to the contrapuntal part a free movement. We will show this to the student in the following example. One adds to the notes of the cantus firmus, generally, only the third or tenth.

Therefore the counterpoint is formed in such a manner, that the progression of thirds and tenths are, as much as possible, covered; for instance:

From this little phrase we could be able to form the following four inversions.

First inversion. The counterpoint is placed in the lower Tenth, the Cantus firmus remains.

Second inversion. The Cantus firmus is placed in the higher Twelfth, the Counterpoint remains.

The third kind of inversion would correspond with the first. The Cantus firmus is placed an octave higher, the Counterpoint a Fifth lower.

The fourth manner of inversion corresponds with the second; the Cantus firmus is placed a Fifth higher, the Counterpoint an Octave lower.

C. f.

264.

Cp.

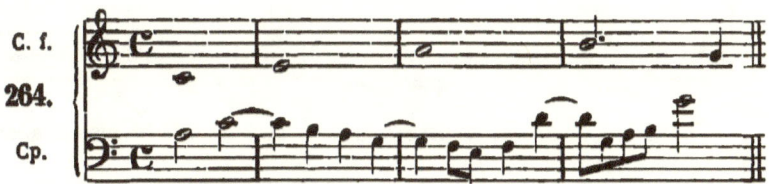

To these two-part phrases can be added one third Free-part, or two Free-parts. We illustrate this in example 260 a, to which we add one free-middle part, and in the inversion 263 a. to which we join two free-lower parts.

The Cantus firmus is placed, an octave lower, to leave room for the free-middle part.

Cp.

260 b.

C. f.

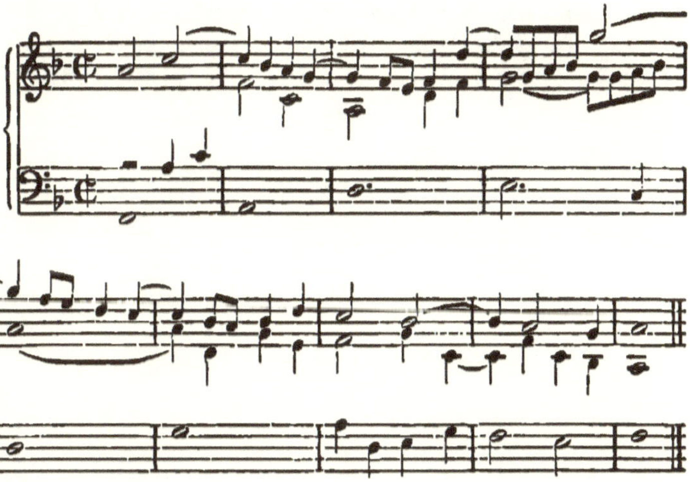

263 b.

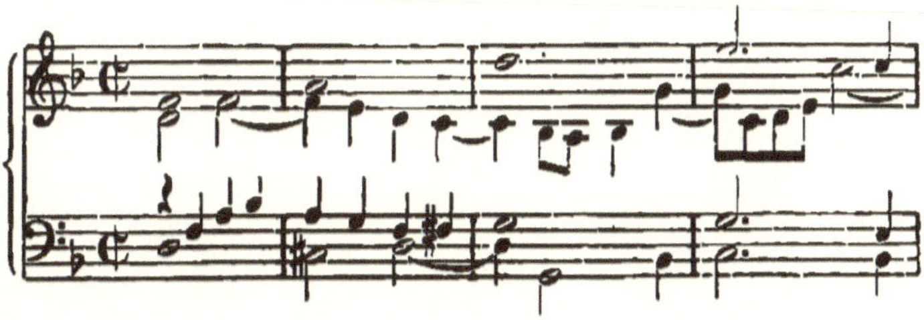

These two free-parts placed a fifth higher will be also available for the inversion of Example 262.

Triple or Quadruple counterpoint in the twelfth does not exist, nor with the double counterpoint in the tenth.

If one wished to add, to a double counterpoint in the twelfth, a third, or a third and fourth part, which have also to be inverted, one would have to follow strictly those rules, given in quadruple counterpoint in the Octave, which we have already worked in Example 253. All those suspensions and passing dissonances, which had to be evaded there, have also to be discarded here. The pupil is recommended to commence his studies in counterpoint in the twelfth, only in two parts, and that, even without especial regard to the two-part phrase. To each example he can add one, or even two free-parts, as we have shown already in example 260 b and 263 b.

PART THIRD.

CHAPTER XI.

Counterpoint in movement of Five, Six, Seven and Eight real Parts.

§ 28. The more conscientiously all the rules and principles of the "serious style" have to be observed, in three and two-part writing, (if the phrase makes any pretention for producing a good effect,) the less strict one need be in movements, containing more than four parts. The more numerous the parts are, the freer can be treated, covered fifths and octaves, doubling of the leading note, the preparation, and resolution of the sevenths, etc.

If in the course of a vocal composition of more than four parts, the phrase be reduced temporally to four, three, or two parts, also all the rules and principles of this style will be again available.

In phrases for seven and eight parts, it is allowed to take the two lowest parts occasionally in octaves or unisons, for instance:

265.

7 *

Let us first elucidate the Five-part movement. Though one could double in this style each of the four parts, still it would be better to write two sopranos, alto, tenor and bass; or perhaps: soprano, alto tenor and two basses. Experience teaches us that in Choral Societies soprano-voices are the most abundant; next to these the basses preponderate above the tenors and altos.

It is therefore better for the effect in sound to divide into two independant parts, those voices that are present in larger numbers, than the less numbering middle parts, which will, as it is, not be heard so much as the extreme parts. For this reason one would place the cantus firmus mostly in the bass, not often in the soprano, and still seldomer into the middle parts. In unequal counterpoint it would be most practical to divide the movement amongst the different parts, and allow it to alternate amongst them. To give movement to *one* part alone, is not advisable in this case. The five-part phrase, composed of first and second soprano, alto, tenor and bass, would produce, in clever hands, an excellent effect by its remarkable fullness of sound.

We give an instance of Five-part style in equal counterpoint, the cantus firmus being in the bass.

266.

Soprano I.

Soprano II.

Alto.

Tenor.

Bass.

C. f.

The same cantus firmus might .be treated in unequal counterpoint in the following manner:

267.

The same cantus firmus with figuration.

268.

§ 29. For the phrase in Six parts the division of two voices each would be most advisable. We write then for I[st] and IInd soprano, alto and tenor, I[st] and IInd bass. The cantus firmus of example 266 would be represented, treated for six parts, in the following manner.

269.

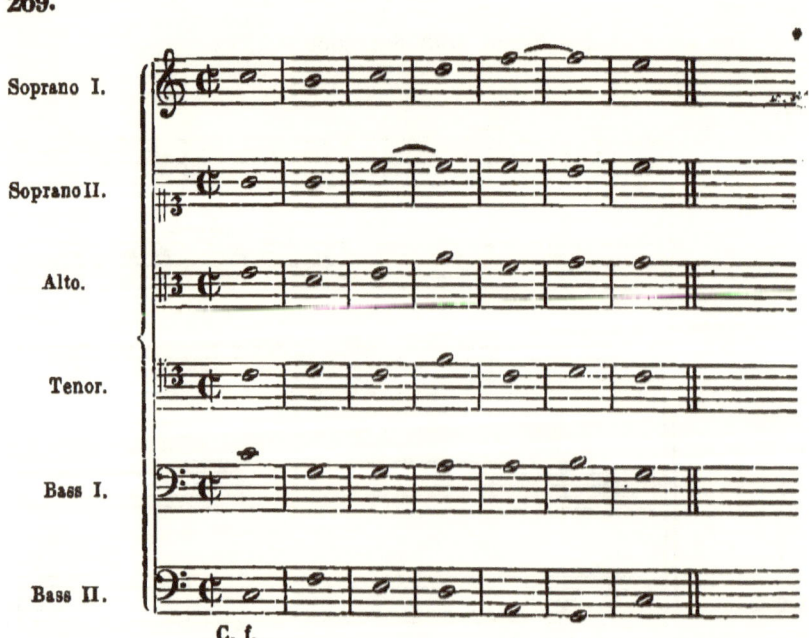

Soprano I.

Soprano II.

Alto.

Tenor.

Bass I.

Bass II.

C. f.

The same cantus firmus with more figuration in the middle-parts:

270a.

It will be perceived, that it is not at all necessary to engage all voices permanently. On the contrary, — the effect will be more beautiful, if now and then some of the voices pause, at proper places, or if the parts enter one after another as in the following example:

270b.

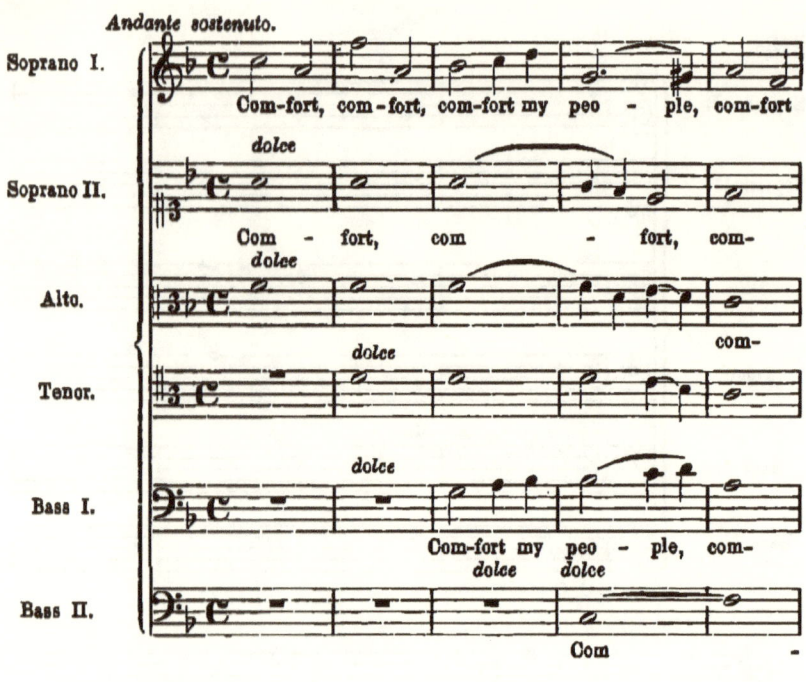

§ 30. In the Seven-part phrase we give the two sopranos and alto the assistance of a male chorus of two tenors and two basses. The above cantus firmus would then appear:

271.

The same cantus firmus with figurated counterpoint:

272.

A suspension over a whole tone resolving upwards, in conjunction with a suspension resolving downwards, (as demonstrated in the alto at NB. last bar but one) is allowable in pure writing (compare Manual of Harmony § 55). Also the hidden octave above the seventh, between the I^st Tenor and the I^st Soprano (at NB. last bar but one) will be permitted at an accumulation of seven parts; as well as all other hidden parallels of octaves and fifths. Also permitted is a succession of a diminished and perfect fifth, (forbidden in four-part writing,) in an ascending direction: ⟨notation⟩. Also octave-parallels in contrary motion are allowable.

The above mentioned hidden octave, above the seventh, could however be evaded easily, by placing in the I^st tenor, as last note but one, *D* instead of *G*.

We give here the same cantus firmus, with more flurried, counterpoint.

273.

Soprano I.

Soprano II.

Alto.

Tenor I.

Tenor II.

Bass I.

Bass II.

One would moreover find it more practicable, at the composition of a movement for seven parts, to imagine it as represented by a double chorus, viz: female chorus with two sopranos and alto and male chorus for four parts. One employs the two choruses, now in alternation, then simultaneously, as treated in the following example:

274.

In phrases for eight parts, every voice can be doubled. One employs however seldomer *one* chorus for eight voices, than *two* choruses for four voices each, which act, partly alternately, partly simultaneously.   One may allow the basses of both choruses to move sometimes in octaves or unisons; the sopranos of both choruses also are led occasionally in unisons; sometimes both choruses are written in such a manner, as to form only *one four-part* chorus.

All these concessions are necessary, on account of the extreme difficulty of leading eight real parts perfectly independently for any length of time. We now give directions, how the student can try to practise this style, firstly in the treatment of our cantus firmus, (hitherto used at all the *poly-part* examples,) showing, for the beginning, an example in equal counterpoint.   The student may carefully observe that none of the parts form parallels of open fifths, or octaves with one another.

275.

The same cantus firmus with figurated counterpoint.

276.

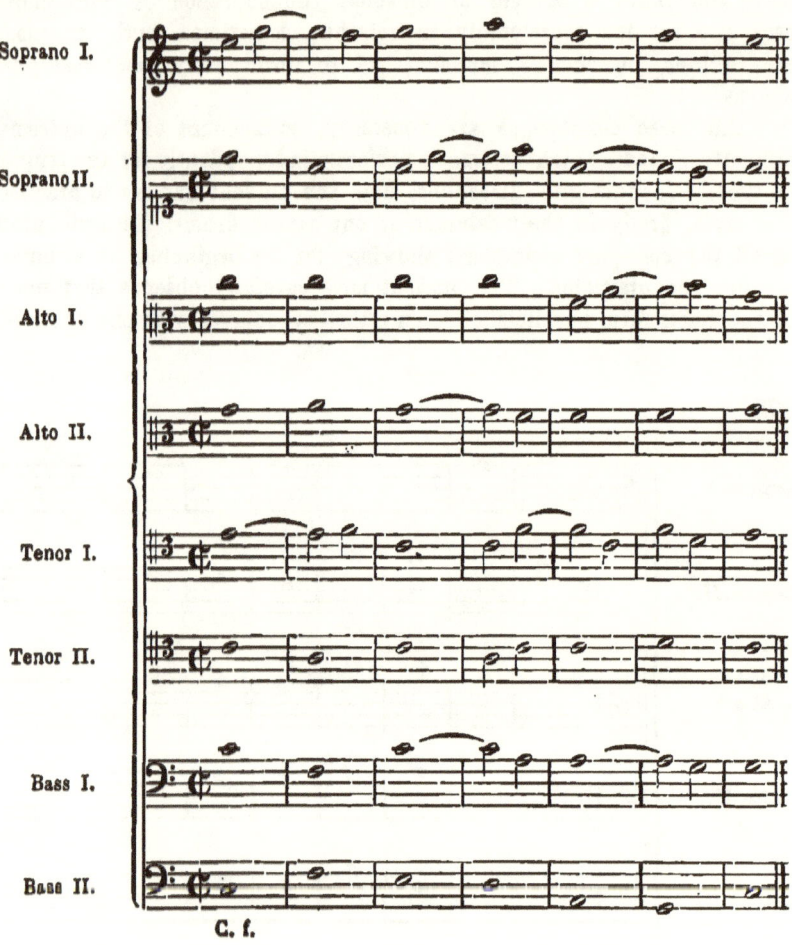

If it be intended to write an ornamented counterpoint in eight-part phrase, especial attention must be given not to place the passing discording notes too near to the consonant notes, in order that the harmony may appear clear at all times. Here follows an example to our often used counterpoint.

277.

In free composition one makes use of all these liberties before
mentioned; allowing, at will, the sopranos or altos, tenors or basses to
go in unisons, or the whole chorus to be in four parts only.

Here is an example of this kind.

278.

One would however attain a much better effect, by writing two choruses in four parts each; here follows an example of such a double chorus.

279.

To compose in this manner for two choruses, one would do well, to treat the voices of the first chorus as I[st] Soprano, I[st] Alto etc. placing the parts of the second chorus as IInd Soprano, IInd Alto lower accordingly.

The student may work for his exercises in five, six, seven and eight parts one or the other cantus firmus of former examples, best suited for this is a bass cantus firmus.

Later on he may endeavour to invent himself, independently, such poly-part movements, and to give them the form of small Motettos. After having now acquired the rules of counterpoint it will be, of the greatest benefit and importance to him, to study industriously the works of the classical authors, such as BACH, HANDEL and others; only then, his studies will lead him to real beneficial results.

Explanatory remarks and hints

for

the treatment of the Exercises in the Treatise on Counterpoint with especial regard to self-instruction.

----

§ 2, page 8. It is evident that the cantus firmus of No. 34 has to be worked in *e*-minor, on account of the *d* in the fifth bar. We give the following example:

A working out of the cantus firmus of No. 35 could be done in the following manner:

NB. The seventh should be led upwards, as the bass takes its natural tone of resolution. (Compare Manual of Harmony § 45.)

§ 10, page 24, Example 85.   A leap into the major seventh
has always to be avoided; it cannot therefore, be used for the preparation of a suspension in a simular manner to the minor or diminished
seventh.   Example 85 *b* shows in the third bar — in a sequence of
several suspensions — the employment of the major seventh, diatonically striking after the octave, used as preparation for the suspension.
As a chord of the seventh with altered fifth, the dissonance of the
major seventh appears less harsh; but then, the suspension becomes
impossible, on account of the altered fifth, which requires dissolving
upwards, (see Manual of Harmony § 48, page 102) which would take,
in this case, the note of resolution, before the suspension itself has
been dissolved.

When two contrapuntal parts move in minims against semi-breves
of the cantus firmus, one can write the exercise in the following
manner:

To § 12, page 30.   For the working out of the cantus firmus
of No. 108, we give a few hints; the counterpoint of the soprano
requires two minims against a semi-breve of the cantus firmus in
the alto.

C. f.

108*a*.

NB. See Manual of Harmony § 53, pag. 131, Ex. 257 b.

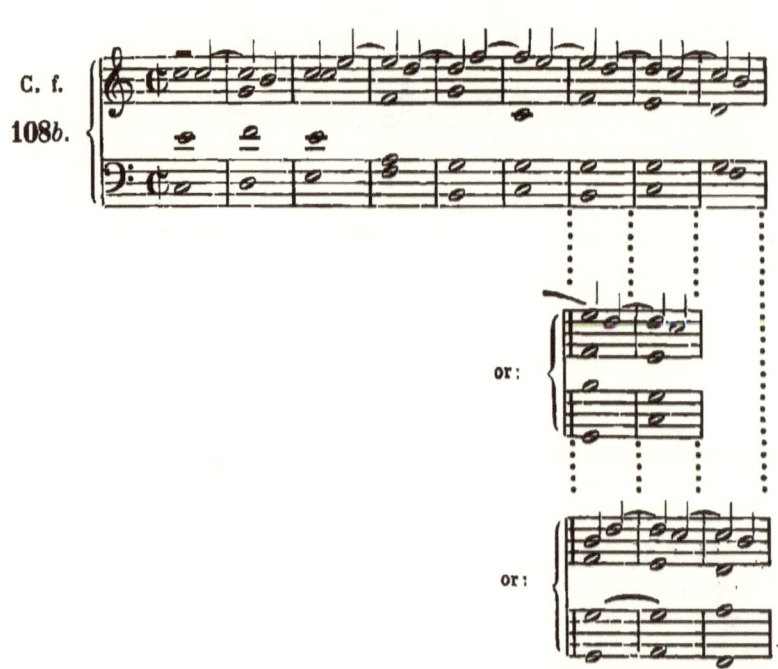

C. f.

108*b*.

or:

or:

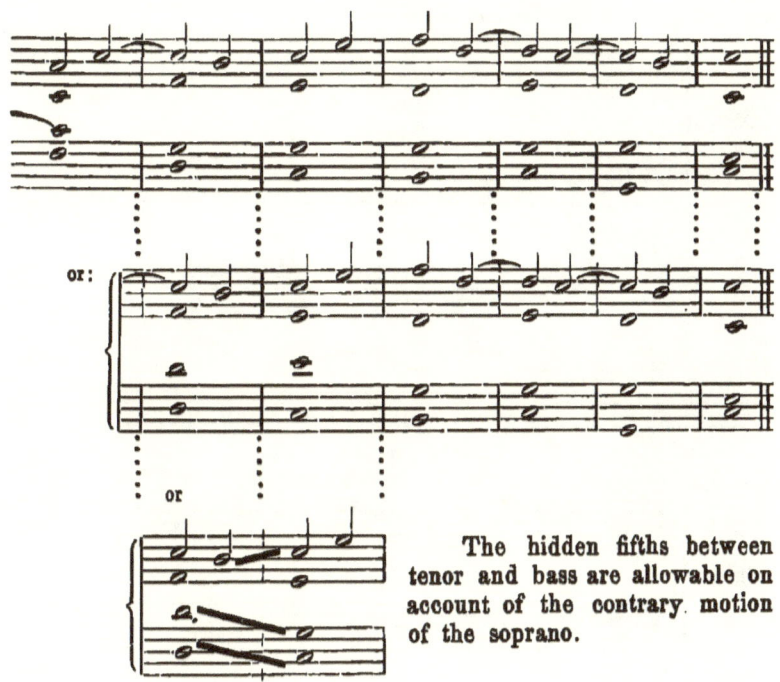

or

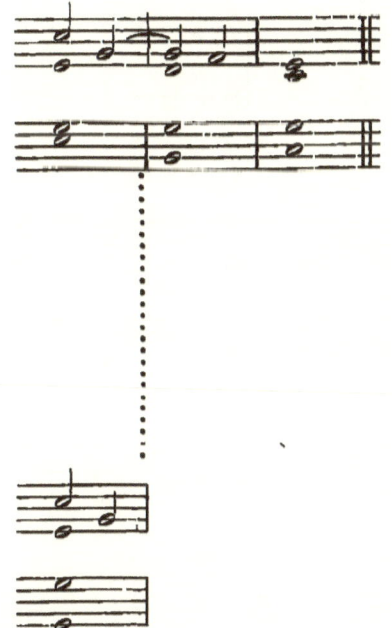

The hidden fifths between tenor and bass are allowable on account of the contrary motion of the soprano.

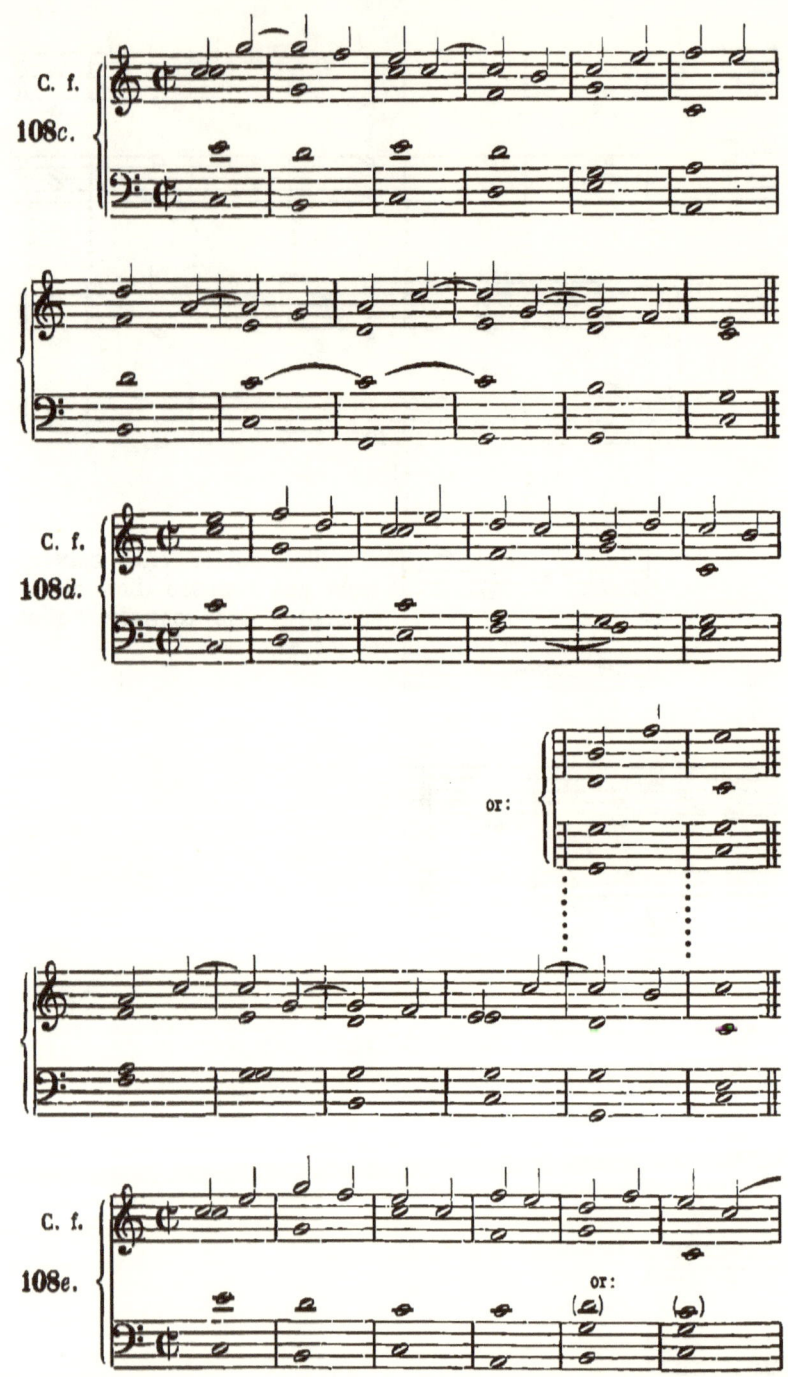

We add a working out of the cantus firmus of No. 100 in form of a sequence.

NB. See Manual of Harmony § 53 pag. 131, Ex. 257 c.

To § 15, Ex. 134 and 135. Either notes of the leap must be harmonious ones.

To § 15, Ex. 136. In this case the most suitable note will be the diatonic passing seventh; less so the passing ninth. In some cases a diatonic progression upwards can be used, after three notes of the same chord, for instance,  A chromatic progression after three notes of the same chord, could only be available in rare cases, for instance:

To § 22, Ex. 201. It will ¡be perceived on observing this example, that in this case the distance of the bass from the alto may amount to two octaves. The careful introduction of the fifth, of the major and minor triad, will be found necessary already here. It will be advisable to introduce this interval, either as passing note (best on an unaccentuated part) or prepared, in order that, at the inversion, the chord of the fourth and sixth, does not enter ill-prepared on an accentuated part. This, only the chord the fourth and sixth, of the chord of the tonic would be allowed to do, as a preparation for the close.

In Ex. 201 in the third bar, the fifth of the dominant (*A*, *C#*, *E*,) appears in the bass on the fourth crotchet. The fundamental note (*A*), struck in the soprano on the first crotchet, and being sustained, serves as a preparation to the fifth. The fifth, itself, is a passing note. In the fourth bar, *D* is sustained in the bass since the first crotchet; on the third crotchet it becomes, (transitorily by the passing note *G* in the soprano) fifth in the chord of the subdominant (*G*, *B*, *D*). The same occurs with the *A* in the alto, in the sixth bar of the same example.

Example 202. The inversion of soprano and bass shows, in the soprano, the fifth of the chords on the second degree, (*e*, *g*, *b*, bar second), the sixth degree, (*b*, *d*, *f#*, bar fourth) and the chord of the dominant, (*A*, *C#*, *E*, bar sixth) as passing notes on the fourth crotchet; fundamental note and third of the respective chords are each time present in other parts. All that has been previously said, concerning preparation and introduction of the fifth remains in force for example 207.

To § 24, Ex. 228. As in the Ex. 201, 202 and 207, we observe also here, that the fifth of the major and minor chords are always carefully prepared. Only in the last bar but one (10), the fifth *F* enters free on the first crotchet. But here it is the fifth of the chord of the Tonic, shortly before the close; and the chord of the fourth and sixth used, is quite in its proper place, and especially well qualified to indicate the approaching conclusion, and to prepare the same. Above all the fundamental note of the chord ($B^\flat$ in the alto) is prepared.

To § 24, Ex. 237. A crossing of parts in an inversion can naturally only occur, when in the original position, the distance between two upper parts becomes greater than an octave, as in the case of Ex. 237, bar 3, between tenor and alto. One will observe from the progression of the soprano, (bar 2) that the altered fifth can be employed advantageously.

To § 25, Ex. 245 *b*. Also here the preparation of the fifth, of the major and minor chords has been observed carefully; only in the eighth bar, we find on the third crotchet, the free entrance of the fifth of the chord on the second degree (c, $e^\flat$, g). The effect is not at all bad in the inversion, (Ex. 245 *c*) because of the fifth being a *chromatic* passing note of no great moment. The latter could have been easily avoided; it has been placed intentionally, in order to descant upon this exceptional case.

To § 26, Ex. 258. If the student places the first 6 notes of the soprano, in the two first bars of this example an octave lower, he will obtain an instance (like No. 249) of an example in double counterpoint in the tenth.

To § 28, 29 and 30. We add still a few more basses, especially adapted for work in more than four-part exercises. The student is meant to treat these, first in equal, afterwards in unequal counterpoint, for 5, 6 and more parts.

INDEX.

––––

Printed by Breitkopf and Härtel, Leipzig.